GREEN CHEMISTRY AND BIODIVERSITY

Principles, Techniques, and Correlations

GREEN CHEMISTRY AND BIODIVERSITY

Principles, Techniques, and Correlations

Edited by

Cristobal N. Aguilar, PhD
Suresh C. Ameta, PhD
A. K. Haghi, PhD

Apple Academic Press Inc.
4164 Lakeshore Road
Burlington ON L7L 1A4
Canada

Apple Academic Press Inc.
1265 Goldenrod Circle NE
Palm Bay, Florida 32905
USA

© 2020 by Apple Academic Press, Inc.

Exclusive worldwide distribution by CRC Press, a member of Taylor & Francis Group

No claim to original U.S. Government works
Printed and bound by CPI Group (UK) Ltd, Croydon, CR0 4YY
International Standard Book Number-13: 978-1-77188-794-6 (Hardcover)
International Standard Book Number-13: 978-0-42920-259-9 (eBook)

All rights reserved. No part of this work may be reprinted or reproduced or utilized in any form or by any electric, mechanical or other means, now known or hereafter invented, including photocopying and recording, or in any information storage or retrieval system, without permission in writing from the publisher or its distributor, except in the case of brief excerpts or quotations for use in reviews or critical articles.

This book contains information obtained from authentic and highly regarded sources. Reprinted material is quoted with permission and sources are indicated. Copyright for individual articles remains with the authors as indicated. A wide variety of references are listed. Reasonable efforts have been made to publish reliable data and information, but the authors, editors, and the publisher cannot assume responsibility for the validity of all materials or the consequences of their use. The authors, editors, and the publisher have attempted to trace the copyright holders of all material reproduced in this publication and apologize to copyright holders if permission to publish in this form has not been obtained. If any copyright material has not been acknowledged, please write and let us know so we may rectify in any future reprint.

Trademark Notice: Registered trademark of products or corporate names are used only for explanation and identification without intent to infringe.

Library and Archives Canada Cataloguing in Publication

Title: Green chemistry and biodiversity : principles, techniques, and correlations / edited by Cristobal N. Aguilar, PhD, Suresh C. Ameta, PhD, A.K. Haghi, PhD.

Names: Aguilar, Cristóbal N, editor. | Ameta, Suresh C., editor. | Haghi, A. K., editor.

Description: Includes bibliographical references and index.

Identifiers: Canadiana (print) 20190184655 | Canadiana (ebook) 20190184671 | ISBN 9781771887946 (hardcover) | ISBN 9780429202599 (ebook)

Subjects: LCSH: Green chemistry. | LCSH: Biodiversity.

Classification: LCC TP155.2.E58 G74 2020 | DDC 660.028/6—dc23

....

CIP data on file with US Library of Congress

....

Apple Academic Press also publishes its books in a variety of electronic formats. Some content that appears in print may not be available in electronic format. For information about Apple Academic Press products, visit our website at **www.appleacademicpress.com** and the CRC Press website at **www.crcpress.com**

About the Editors

Cristobal N. Aguilar, PhD
Full Professor and Dean, School of Chemistry, Autonomous University of Coahuila, México

Cristobal N. Aguilar, PhD, is Full Professor and Dean of the School of Chemistry at the Autonomous University of Coahuila, México. Dr. Aguilar has published more than 160 papers published in indexed journals, more than 40 articles in Mexican journals, as well as 16 book chapters, eight Mexican books, four international books, 34 proceedings, and more than 250 contributions in scientific meetings. Professor Aguilar is a member of the National System of Researchers of Mexico (SNI) and has received several prizes and awards, the most important are the National Prize of Research (2010) of the Mexican Academy of Sciences, the "Carlos Casas Campillo 2008" prize of the Mexican Society of Biotechnology and Bioengineering, the National Prize AgroBio-2005, and the Mexican Prize in Food Science and Technology from CONACYT-Coca Cola México in 2003. He is also a member of the Mexican Academy of Science, the International Bioprocessing Association (IFIBiop), and several other scientific societies and associations. Dr. Aguilar has developed more than 21 research projects, including six international exchange projects. He has been the advisor of 18 PhD theses, 25 MSc theses, and 50 BSc theses. He became a Chemist at the Autonomous University of Coahuila, México, and earned his MSc degree in Food Science and Biotechnology at the Autonomous University of Chihuahua, México. His PhD degree in Fermentation Biotechnology was awarded by the Autonomous University of Metropolitana, Mexico. Dr. Aguilar also performed postdoctoral work at the Department of Biotechnology and Molecular Microbiology at Research Institute for Development (IRD) in Marseille, France.

Suresh C. Ameta, PhD

Dean, Faculty of Science, PAHER University, Udaipur, India

Suresh C. Ameta, PhD, is currently Dean, Faculty of Science at PAHER University, Udaipur, India. He has served as Professor and Head of the Department of Chemistry at North Gujarat University Patan and at M. L. Sukhadia University, Udaipur, and as Head of the Department of Polymer Science. He also served as Dean of Postgraduate Studies. Prof. Ameta has held the position of President of the Indian Chemical Society, Kotkata, and is now a life-long Vice President. He was awarded a number of prestigious awards during his career, such as national prizes twice for writing chemistry books in Hindi. He also received the Prof. M. N. Desai Award (2004), the Prof. W. U. Malik Award (2008), the National Teacher Award (2011), the Prof. G. V. Bakore Award (2007), a Life-Time Achievement Award by the Indian Chemical Society (2011) as well as the Indian Council of Chemist (2015), etc. He has successfully guided 81 PhD students. Having more than 350 research publications to his credit in journals of national and international repute, he is also the author of many undergraduate- and postgraduate-level books. He has published three books with Apple Academic Press: *Chemical Applications of Symmetry and Group Theory; Microwave-Assisted Organic Synthesis*; and *Green Chemistry: Fundamentals and Applications*; and two with Taylor and Francis: *Solar Energy Conversion and Storage* and *Photocatalysis*. He has also written chapters in books published by several other international publishers. Prof. Ameta has delivered lectures and chaired sessions at national conferences and is a reviewer for number of international journals. In addition, he has completed five major research projects for different funding agencies, such as DST, UGC, CSIR, and Ministry of Energy, Govt. of India.

About the Editors

A. K. Haghi, PhD
Professor Emeritus of Engineering Sciences, Former Editor-in-Chief, *International Journal of Chemoinformatics* and *Chemical Engineering & Polymers Research Journal*; Member, Canadian Research and Development Center of Sciences and Cultures (CRDCSC), Canada

A. K. Haghi, PhD, is the author and editor of 165 books, as well as 1000 published papers in various journals and conference proceedings. Dr. Haghi has received several grants, consulted for a number of major corporations, and is a frequent speaker to national and international audiences. Since 1983, he served as professor at several universities. He is the former Editor-in-Chief of the *International Journal of Chemoinformatics and Chemical Engineering* and *Polymers Research Journal* and is on the editorial boards of many international journals. He is also a member of the Canadian Research and Development Center of Sciences and Cultures (CRDCSC), Montreal, Quebec, Canada.

Contents

Contributors .... *xi*

Abbreviations .... *xv*

Preface .... *xvii*

1. **Photodegradation of 2-Nitrophenol, an Endocrine Disruptor, Using TiO$_2$ Nanospheres/SnO$_2$ Quantum Dots** 1

   Jayesh Bhatt, Kanchan Kumari Jat, Avinash K. Rai, Rakshit Ameta, and Suresh C. Ameta

2. **Biodiversity as a Source of Drugs: *Cordia, Echinacea, Tabernaemontana*, and *Aloe*** 9

   Francisco Torrens and Gloria Castellano

3. **Biodiversity: Loss and Conservation** 23

   Anamika Singh and Rajeev Singh

4. ***Aegle marmelos*: Nature's Gift for Human Beings** 35

   Rajeev Singh and Anamika Singh

5. **Seed-Growth Method for the Synthesis of Metal Nanoparticles** 47

   Lavanya Tandon, Divya Mandial, Rajpreet Kaur, and Poonam Khullar

6. **Soil-Protecting Functions of Medicinal Plants: Meadow and Field Weeds** 65

   Rafail A. Afanas'ev, Genrietta E. Merzlaya, and Michail O. Smirnov

7. **Glycosylation of Polyphenols in Tannin-Rich Extracts from *Euphorbia antisyphilitica, Jatropha dioica*, and *Larrea tridentata*** 81

   Janeth Ventura-Sobrevilla, Gerardo Gutiérrez-Sánchez, Carl Bergmann, Parastoo Azadi, Daniel Boone-Villa, Raul Rodriguez, and Cristóbal N. Aguilar

8. **Analysis and Quantification of *Larrea tridentata* Polyphenols Obtained by Reflux and Ultrasound-Assisted Extraction** 97

   Karina Cruz-Aldaco, Daniela Sánchez-Aldana, Salvador Ortega-Hernández, Guadalupe Cárdenas-Félix, Antonio Aguilera-Carbo, Juan Alberto Ascasio-Valdes, Raúl Rodriguez-Herrera, and Cristóbal Noé Aguilar

Contents

9. Properties and Applications of the Phytochemical: Ellagic Acid (4,4,5,5,6,6-hexahydroxydiphenic acid-2,6,2,6-dilactone) 111

René Díaz-Herrera, Pedro Aguilar-Zarate, Juan A. Ascacio-Valdes, Leonardo Sepúlveda-Torre, Juan Buenrostro-Figueroa, Monica L. Chavez-Gonzalez, Janeth Ventura, and Cristóbal N. Aguilar

10. Antioxidative Properties of *Punica granatum*, *Peganum harmala*, *Dianthus caryophyllus*, and *Vitis vinifera* Extracts Against Free Radicals 135

Katarína Valachová, Elsayed E. Hafez, Milan Nagy, and Ladislav Šoltés

11. Flavonoids for Designing Metal Nanoparticles and Their Applications 153

Divya Mandial, Rajpreet Kaur, and Poonam Khullar

12. pH and Temperature Factor Affecting Curcumin Properties and Its Bioapplicability 169

Rajpreet Kaur, Divya Mandial, Lavanya Tandon, and Poonam Khullar

13. Integrated Water Resource Management and Nanotechnology Applications in Water Purification: A Critical Overview 187

Sukanchan Palit

14. Precision Personalized Medicine from Theory to Practice: Cancer... 209

Francisco Torrens and Gloria Castellano

15. Design, Synthesis, and Studies of Novel Piperidine Substituted Triazine Derivatives as Potential Anti-Inflammatory and Antimicrobial Agents 243

Ravindra S. Shnide

16. Metaphors That Made History: Reflections on Philosophy/Science/DNA 267

Francisco Torrens and Gloria Castellano

Index... *281*

Contributors

Rafail A. Afanas'ev
Pryanishnikov All-Russian Scientific Research Institute of Agrochemistry, d. 31A, Pryanishnikova Street, Moscow 127550, Russia

Cristóbal Noé Aguilar
Food Research Department, School of Chemistry, Universidad Autónoma de Coahuila, Blvd. Venustiano Carranza S/N Col. República Oriente, 25280 Saltillo, Coahuila, México

Pedro Aguilar-Zarate
Tecnológico Nacional de México, Instituto Tecnológico de Ciudad Valles, Ciudad Valles, SLP, México

Antonio Aguilera-Carbo
Department of Animal Nutrition, Universidad Autónoma Agraria Antonio Narro, Calzada Antonio Narro 1923, Col. Buenavista, 25315 Saltillo, Coahuila, México

Rakshit Ameta
Department of Chemistry, PAHER University, Udaipur 313003, Rajasthan, India
Department of Chemistry, Faculty of Science, J. R. N. Rajasthan Vidyapeeth (Deemed-to-be University), Udaipur 313002, Rajasthan, India

Suresh C. Ameta
Department of Chemistry, PAHER University, Udaipur 313003, Rajasthan, India

Juan Alberto Ascasio-Valdes
Food Research Department, School of Chemistry, Universidad Autónoma de Coahuila, Blvd. Venustiano Carranza S/N Col. República Oriente, 25280 Saltillo, Coahuila, México

Parastoo Azadi
Complex Carbohydrate Research Center, The University of Georgia, Athens, GA 33602-4712, USA

Carl Bergmann
Complex Carbohydrate Research Center, The University of Georgia, Athens, GA 33602-4712, USA

Jayesh Bhatt
Department of Chemistry, PAHER University, Udaipur 313003, Rajasthan, India

Daniel Boone-Villa
Food Research Department, School of Chemistry, Universidad Autónoma de Coahuila. PO Box 252, 25000, Saltillo, Coahuila, México

Juan Buenrostro-Figueroa
Research Center for Food and Development A.C., Delicias City, Chihuahua, México

Guadalupe Cárdenas-Félix
Innovation and Technological Development Center, BAFAR Group, Km 7.5 Carretera a Cuauhtémoc S/N Col. Las ánimas, 31450 Chihuahua, Chihuahua, Mexico

Gloria Castellano
Departamento de Ciencias Experimentales y Matemáticas, Facultad de Veterinaria y Ciencias Experimentales, Universidad Católica de Valencia San Vicente Mártir, Guillem de Castro-94, E-46001 València, Spain

Monica L. Chavez-Gonzalez
Bioprocesses and Bioproducts Group, DIA-UAdeC Food Research Department, School of Chemistry, Universidad Autónoma de Coahuila, Saltillo 25280, México

Karina Cruz-Aldaco
Food Research Department, School of Chemistry, Universidad Autónoma de Coahuila, Blvd. Venustiano Carranza S/N Col. República Oriente, 25280 Saltillo, Coahuila, México

René Díaz-Herrera
Bioprocesses and Bioproducts Group, DIA-UAdeC Food Research Department, School of Chemistry, Universidad Autónoma de Coahuila, Saltillo 25280, México

Gerardo Gutiérrez-Sánchez
Complex Carbohydrate Research Center, The University of Georgia, Athens, GA 33602-4712, USA

Elsayed E. Hafez
Plant Protection and Biomolecular Diagnosis Department, Arid Lands Cultivation Research Institute, City of Scientific Research and Technological Applications, New Borg El-Arab City, Alexandria 21934, Egypt

Rajpreet Kaur
Department of Chemistry, B.B.K. D.A.V. College for Women, Amritsar 143005, Punjab, India

Poonam Khullar
Department of Chemistry, B.B.K. D.A.V. College for Women, Amritsar 143005, Punjab, India

Kanchan Kumari Jat
Department of Chemistry, PAHER University, Udaipur 313003, Rajasthan, India
Department of Chemistry, Mohan Lal Sukhadia University, Udaipur 313002, Rajasthan, India

Divya Mandial
Department of Chemistry, B.B.K. D.A.V. College for Women, Amritsar 143005, Punjab, India

Genrietta E. Merzlaya
Pryanishnikov All-Russian Scientific Research Institute of Agrochemistry, d. 31A, Pryanishnikova Street, Moscow 127550, Russia

Milan Nagy
Department of Pharmacognosy and Botany, Faculty of Pharmacy, Comenius University, Bratislava, Slovakia

Salvador Ortega-Hernández
Innovation and Technological Development Center, BAFAR Group, Km 7.5 Carretera a Cuauhtémoc S/N Col. Las ánimas, 31450 Chihuahua, Chihuahua, Mexico

Sukanchan Palit
43, Judges Bagan, Post Office Haridevpur, Kolkata 700082, India

Avinash K. Rai
Department of Chemistry, PAHER University, Udaipur 313003, Rajasthan, India

Contributors

Raúl Rodriguez-Herrera
Food Research Department, School of Chemistry, Universidad Autónoma de Coahuila, Blvd. Venustiano Carranza S/N Col. República Oriente, 25280 Saltillo, Coahuila, México

Daniela Sánchez-Aldana
Innovation and Technological Development Center, BAFAR Group, Km 7.5 Carretera a Cuauhtémoc S/N Col. Las ánimas, 31450 Chihuahua, Chihuahua, Mexico

Leonardo Sepúlveda-Torre
Bioprocesses and Bioproducts Group, DIA-UAdeC Food Research Department, School of Chemistry, Universidad Autónoma de Coahuila, Saltillo 25280, México

Ravindra S. Shnide
Department of Chemistry and Industrial Chemistry, Dayanand Science College, Latur 413512, Maharashtra, India

Anamika Singh
Department of Botany, Maitreyi College, University of Delhi, Delhi, India
Department of Environmental Studies, Satyawati College, University of Delhi, Delhi, India

Rajeev Singh
Department of Botany, Maitreyi College, University of Delhi, Delhi, India
Department of Environmental Studies, Satyawati College, University of Delhi, Delhi, India

Michail O. Smirnov
Pryanishnikov All-Russian Scientific Research Institute of Agrochemistry, d. 31A, Pryanishnikova Street, Moscow 127550, Russia

Ladislav Šoltés
Center of Experimental Medicine, Institute of Experimental Pharmacology and Toxicology, Slovak Academy of Sciences, Bratislava, Slovakia

Lavanya Tandon
Department of Chemistry, B.B.K. D.A.V. College for Women, Amritsar 143005, Punjab, India

Francisco Torrens
Institut Universitari de Ciència Molecular, Universitat de València, Edifici d'Instituts de Paterna, P. O. Box 22085, E-46071 València, Spain

Katarína Valachová
Center of Experimental Medicine, Institute of Experimental Pharmacology and Toxicology, Slovak Academy of Sciences, Bratislava, Slovakia

Janeth Ventura-Sobrevilla
Food Research Department, School of Chemistry, Universidad Autónoma de Coahuila, PO Box 252, 25000 Saltillo, Coahuila, México

Abbreviations

2D	two-dimensional
ACN	acetonitrile
ACS	American Chemical Society
AIDS	acquired immunodeficiency syndrome
AuNPs	gold nanoparticles
CNS	central nervous system
CTAB	cetyltrimethylammonium Bromide bromide
DLS	dynamic light scattering
DPPH	1,1-diphenyl-2-picrylhydrazyl
EA	ellagic acid
EAF	*E. antisyphilitica* aqueous fraction
EAH	ellagitannin-acyl hydrolase
EC	*E. antisyphilitica* crude
EDCs	endocrine-disrupting chemicals
EDL	electronic double layer
EDX	energy dispersive X-ray
EEF	*E. antisyphilitica* ethanolic fraction
ETs	ellagitannins
FCC	face-centered cubic
GC–MS	gas chromatography–mass spectrometry
GIS	geographical information systems
HDL	high-density lipoprotein
HGP	Human Genome Project
HHDP	hexahydroxydiphenic acid
HIV	human immunodeficiency virus
JAF	*J. dioica* aqueous fraction
JC	*J. dioica* crude
JEF	*J. dioica* ethanolic fraction
LAF	*L. tridentata* aqueous fraction
LC	*L. tridentata* crude
LDL	low-density lipoprotein
LEF	*L. tridentata* ethanolic fraction
LPS	lip polysaccharide

MDCs	metabolism disrupting chemicals
MIC	minimum inhibitory concentration
MR	multidrug resistance
NCs	nanocomposites
NDGA	nordihydroguaiaretic acid
NMs	nanomaterials
NPs	natural products
PTE	periodic table of the elements
QDs	quantum dots
RAM	radial arm maze
SAED	selected area electron diffraction
SAR	structure–activity relationships
SDHM	solid dispersions by hot melt method
SERS	surface enhance Raman scattering
SSC	solid state culture
STLs	sesquiterpene lactones
TEM	transmission electron microscope
TFA	trifluoro acetic acid
XPS	X-ray photoelectron spectroscopy

Preface

The objective of this book is to report new approaches to designing chemicals and chemical transformations that are beneficial for human health and the environment, which is an area that continues to emerge as an important field of study. Biodiversity is actually a natural wealth of the planet and fulfills all the requirements of any organism. Due to excess utilization of natural resources, loss of biodiversity occurs, so there is a need for conservation and protection of environment.

This volume is a collection of innovative research on the development of alternative sustainable technologies. It takes a broad view of the subject and integrates a wide variety of approaches.

This book introduces the topics with an overview of key concepts and with the latest research and applications, providing both the breadth and depth researchers need.

By focusing on both the interdisciplinary applications of green chemistry and the biodiversity, this book manages to present two key messages in a manner where they reinforce each other. It provides a single and concise reference for chemists, instructors, and students for learning about green chemistry and its great and ever-expanding number of applications.

This book:

- Aims to elucidate the overlap between the green chemistry and biological sciences.
- Demonstrates how green chemistry can contribute to our understanding of biodiversity with the ultimate goal of benefiting from our preserving nature.
- Includes chapters on many aspects of biodiversity and green chemistry at the molecular and macromolecular levels.
- Provides real-world examples of key issues.

CHAPTER 1

Photodegradation of 2-Nitrophenol, an Endocrine Disruptor, Using TiO$_2$ Nanospheres/SnO$_2$ Quantum Dots

JAYESH BHATT[1,*], KANCHAN KUMARI JAT[2], AVINASH K. RAI[1], RAKSHIT AMETA[3], and SURESH C. AMETA[1]

[1]*Department of Chemistry, PAHER University, Udaipur 313003, Rajasthan, India*

[2]*Department of Chemistry, Mohan Lal Sukhadia University, Udaipur 313002, Rajasthan, India*

[3]*Department of Chemistry, Faculty of Science, J. R. N. Rajasthan Vidyapeeth (Deemed-to-be University), Udaipur 313002, Rajasthan, India*

*Corresponding author. E-mail: jay.bht@gmail.com

ABSTRACT

SnO$_2$ quantum dots (QDs) decorated on TiO$_2$ nanospheres have been synthesized and used as photocatalyst in photodegradation of 2-nitrophenol. The photocatalytic behavior of the as-synthesized samples shows a more effective degradation of 2-nitrophenol as compared to pure TiO$_2$ nanospheres. This study offers for potential applications of QDs decorated semiconducting titanium dioxide in treatment of polluted water by degrading pollutants.

1.1 INTRODUCTION

In the last few decades, humans have been showing ignorance toward conscious usage of water and its contamination. It has increased ever increasing scarcity of potable water. Anthropogenic contaminants are increasing exponentially. Most of these are highly toxic to living beings. Endocrine disruptors are chemicals that can interfere with endocrine (or hormone) systems. Most of them are toxic, xenogenic, and carcinogenic in nature. Environmental contamination by these toxic chemicals is causing serious global problems. Various methods and techniques such as biological methods, physical methods, chemical methods, and so on are developed to remove these pollutants. Every method has its own pros and cons. Photocatalytic wastewater treatment has gained much importance recently. Cost-effectiveness, eco-friendly nature, and applicability on large scale for different kind of pollutants are the key factors. Photocatalysts used at present are basically semiconductors like TiO_2, ZnO, CdS, and so on. Their composite may prove to be more beneficial for efficient decomposition of EDCs.

With growing population and increased demands from agriculture and industries, researchers have estimated that portable water crisis is likely to worsen much in coming years. Photocatalysis has emerged as an advanced oxidation process (a green chemical approach) for such a treatment. The term endocrine disruptor was coined in 1991 at the Wingspread Conference Center in Wisconsin. One of the early papers on the phenomenon was studied by Colborn et al.[1] on the developmental effects of endocrine disrupting chemicals in wildlife and humans. Endocrine disruptors are chemicals that at certain doses can interfere with the body's endocrine system (also named hormone systems) in human, mammals, birds, fish, and many other types of living organisms. Endocrine-disrupting chemicals (EDCs) with low dose matters, because of ubiquitous exposure and persistence of biological effects, have wide range of adverse health effects.[2] Impaired brain development, infertility, birth defects, obesity, diabetes, and endometriosis are some of many health problems linked to exposure to EDCs. Sifakis et al.[3] showed that EDCs on human exposure are affecting male and female reproductive systems. Mallozzi et al.[4] studied the carcinogenic effects of EDCs while Braun[5] suggested that pre- and/or early postnatal exposure to some EDCs may increase the risk of overweight or obesity and neurobehavioral disorders during childhood.

Mimanto et al.[6] discussed molecular and cellular mechanisms by which metabolism disrupting chemicals (MDCs) impair energy homeostasis.

Sidorkiewicz et al.[7] showed four modes of action of EDCs on male fertility. Poland and Knutson[8] examined the biochemical and toxic responses produced by halogenated aromatic hydrocarbons and showed that these compounds bind to a cellular receptor and evoke a sustained pleiotropic response.

Exposure to EDCs in nature is a basic cause of worry with unclear long-term negative impacts. EDCs are discharged into nature by people, animals, and industry through sewage, soil, surface water, and groundwater. So, degradation and decomposition of EDCs become highly important. Monitoring of endocrine disrupting chemicals (phenols and phthalates) in South Africa was conducted by Olujimi et al.[9]

In photocatalysis, efficient photocatalysts, nanomaterials, and novel photocatalysts in various shapes and sizes, that is, nanoflakes, nanotubes, nanowires, nanosheets, nanoflowers, nanorods, nanodisks, nanocubes, nanocoreshells, nanobowl, nanocapsules, nanopillars, and so on, are used with wide range of variations in their composites to increase their efficiency in decomposition of EDCs. TiO_2 has been the most studied photocatalyst. Lee et al.[10] developed TiO_2, which is decorated with SnO_2 quantum dots (QDs) for controlling the band structure of TiO_2 photocatalyst. It is a simple and fast surfactant free method for the development and syntheses of SnO_2 QDs that are size-controllable. The decoration of the SnO_2 QDs on the TiO_2 nanoparticles leads to enhance the separation of charges and aids generation of extra oxidizing species and radicals; as a result, boosting the performance of the TiO_2 nanoparticles. Singhal et al.[11] carried out photocatalytic degradation of cetylpyridinium chloride over titanium dioxide powder. Sood et al.[12] used TiO_2 QDs to carry out photocatalytic degradation of indigo carmine dye. However, use of QDs or its composites as photocatalysts still remained unveiled. Kaur et al.[13] reported sunlight-driven photocatalytic degradation of ketorolac tromethamine using TiO_2 QDs.

When a composite is prepared with a wide bandgap semiconductor, which has more positive conduction band edge than that of the smaller bandgap semiconductor, then the electrons from smaller bandgap semiconductor can be injected into the larger bandgap semiconductor, so that coupling can effectively prevent the recombination.[14,15]

4   Green Chemistry and Biodiversity: Principles, Techniques, and Correlations

1.2 EXPERIMENTAL

1.2.1 PREPARATION OF SnO_2 QD

2.5 g of tin chloride pentahydrate ($SnCl_4.5H_2O$; Sigma-Aldrich) was dissolved in 125 mL of distilled water. After stirring for 10 min, 5.2 mL of hydrazine was added to this solution. This mixed solution was stirred continuously for another 30 min at room temperature. Then, the solution was maintained at 100°C for 18 h. SnO_2 QDs were obtained by centrifugation washed with ethanol and water several times. The final product was obtained after drying in oven at 60°C for 12 h.[16]

1.2.2 PREPARATION OF SnO_2 QD@TiO_2 FOR PHOTODEGRADATION OF 2-NITROPHENOL

0.05 g of P25 TiO_2 (Sigma-Aldrich) was added into 10 mL of ethanol (95 vol%) under stirring for 15 min, followed by addition of 300 μL of ethanolic suspension of SnO_2 QD (containing 2 mg of SnO_2 QD). The final mixture was stirred for 30 min at room temperature. The product SnO_2 QD@TiO_2 was then collected by centrifugation and rinsed with ethanol several times before being dried in vacuum oven set at 80°C overnight to afford the final product.[10]

1.2.3 THE DEGRADATION PROFILE OF 2-NITROPHENOL

The photocatalytic activity of the catalyst was evaluated by measuring the rate of degradation of 2-nitrophenol (2NP). A stock solution of (1.0×10^{-3} M) was prepared by dissolving 0.0139 g of 2NP in 100 mL doubly distilled water. pH of the dye solution was measured by a digital pH meter (Systronics model 335), and the desired pH of the solution was adjusted by the addition of standard 0.1 N sulfuric acid and 0.1 N sodium hydroxide solutions. The reaction mixture containing 0.10 g photocatalyst was exposed to a 200 W tungsten lamp, and about 3 mL aliquot was taken out every 10 min. Absorbance (A) was measured at λ_{max} 280 nm. A water filter was used to cut off thermal radiations. The intensity of light was varied by changing the distance between the light source and reaction mixture, and it was measured by Suryamapi (CEL model SM 201). The absorbance

Photodegradation of 2-Nitrophenol

of the solution at various time intervals was measured with the help of spectrophotometer (Model UV-1700 Pharmaspec).

This solution was placed in equal amounts in four beakers. In the first beaker, only 2-nitrophenol solution was taken, which was kept in dark. In the second beaker, the 2-nitrophenol solution was exposed to light. In the remaining two beakers, 2-nitrophenol solution with 0.10 g TiO_2/SnO_2 was taken and kept in dark and exposed to the light, respectively. After keeping these beakers for 3–4 h, the absorbance of the solution in each beaker was measured with the help of a spectrophotometer. It was found that the absorbance of solutions of first three beakers remained almost constant, while the solution of fourth beaker had a decrease in initial value of absorbance. From these observations, it is clear that this reaction requires presence of both; the light as well as semiconductor composite TiO_2/SnO_2 or TiO_2 alone. Hence, this reaction is a photocatalytic reaction in nature.

A solution of 2.5×10^{-4} M of 2-nitrophenol was prepared in doubly distilled water and 0.12 g of TiO_2/SnO_2 and TiO_2 was added to it. The pH of the reaction mixture was adjusted to 6.0 and then this solution was exposed to a 200 W tungsten lamp at 60 mWcm^{-2}. A decrease was observed in the absorbance of 2-nitrophenol solution with increasing time of exposure. The photocatalytic degradation of the 2-nitrophenol was studied at λ_{max} = 280 nm. A plot of $1 + \log A$ versus time was found to be linear, which indicates that the photocatalytic degradation of 2-nitrophenol follows pseudo-first order kinetic.

The rate constant was calculated by using the expression:

$$k = 2.303 \times \text{slope.} \qquad (1.1)$$

A typical run is given in Table 1.1 and is graphically represented in Figure 1.1.

TABLE 1.1 A Typical Run.

Time (min)	Absorbance in presence of TiO_2 (A)	$1 + \log A$	Absorbance in presence of TiO_2/SnO_2 (A)	$1 + \log A$
0	1.11	1.0453	1.11	1.0453
30	1.06	1.0253	1.048	1.0204
60	1.026	1.0111	0.99	0.9956

TABLE 1.1 *(Continued)*

Time (min)	Absorbance in presence of TiO$_2$ (A)	1 + log A	Absorbance in presence of TiO$_2$/SnO$_2$ (A)	1 + log A
90	0.979	0.9908	0.931	0.9689
120	0.936	0.9713	0.879	0.9440
150	0.889	0.9489	0.826	0.9170
180	0.8575	0.9332	0.762	0.8817

pH = 6.0; [2-NP] = 2.40 × 10^{-4} M; semiconductor = 0.12 g; light intensity = 60.0 mWcm^{-2}.

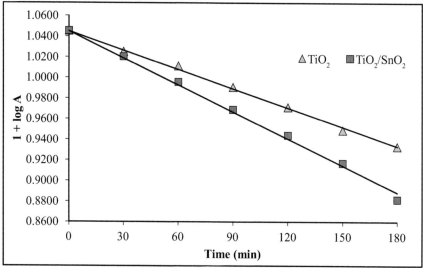

FIGURE 1.1 A comparative typical run of TiO$_2$ and TiO$_2$/SnO$_2$.

The rate constant calculated in presence of TiO$_2$ was found to be 2.39 × 10^{-5} s^{-1} whereas rate constant in presence of TiO$_2$/SnO$_2$ was 3.49 × 10^{-5} s^{-1}, which is almost 46.60% higher, composite. It may be concluded that SnO$_2$ decoration of nano-TiO$_2$ increases its photocatalytic efficiency in visible range for degradation of 2-nitrophenol.

The TiO$_2$/SnO$_2$ composites have been successfully prepared and employed as photocatalyst in photodegradation of 2-nitrophenol. The UV–vis light-driven photocatalytic properties of the as-obtained photocatalysts were investigated systematically with variation of all operational parameters. Due to the special structure and the synergistic effect of the composite, TiO_2/SnO_2 composite exhibited a broad absorption in the UV region and as a result, excellent photocatalytic activity. It provides a facile method to prepare TiO_2/SnO_2 composite, which can be a better choice as a photocatalyst in treatment of wastewater containing 2-nitrophenol.

KEYWORDS

- quantum dots
- endocrine disrupting chemicals
- waste water treatment
- TiO_2
- SnO_2

REFERENCES

1. Colborn, T.; vom Saal, F. S.; Soto, A. M. Developmental Effects of Endocrine-disrupting Chemicals in Wildlife and Humans. *Environ. Health Perspect.* **1993,** *101* (5), 378–384.
2. National Institutes of Environmental Health Sciences (NIEHS), *Endocrine Disruptors*; National Institutes of Health (NIH) & U.S. Department of Health & Human Services, 2010. http://www.niehs.nih.gov/health/topics/agents/endocrine.
3. Sifakis, S.; Androutsopoulos, V. P.; Tsatsakis, A. M.; Spandidos, D. A. Human Exposure to Endocrine Disrupting Chemicals: Effects on the Male and Female Reproductive System. *Environ. Toxicol. Pharmacol.* **2017,** *51*, 56–70.
4. Mallozzi, M.; Leone, C.; Manurita, F.; Bellati F.; Caserta, D. Endocrine Disrupting Chemicals and Endometrial Cancer: An Overview of Recent Laboratory Evidence and Epidemiological Studies. *Int. J. Environ. Res. Public Health.* **2017,** *14* (3), 334. DOI:10.3390/Ijerph14030334.
5. Braun, J. M. Early-Life Exposure to EDCs: Role in Childhood Obesity and Neurodevelopment. *Nat. Rev. Endocrinol.* **2017,** *13*, 161–173.
6. Mimanto, M. S.; Nadal, A.; Sargis, R. M. Polluted Pathways: Mechanisms of Metabolic Disruption by Endocrine Disrupting Chemicals. *Curr. Environ. Health Rpt.* **2017,** *4* (2), 208–222. DOI:10.1007/S40572-017-0137-0.
7. Sidorkiewicz, I.; Zareba, K.; Wołczynski, S.; Czerniecki, J. Endocrine-disrupting Chemicals-mechanisms of Action on Male Reproductive System. *Toxicol. Ind. Health* **2017,** *33* (7), 601–609. DOI: 10.1177/0748233717695160.

8. Poland, A.; Knutson, J. C. 2, 3, 7, 8-Tetrachlorodibenzop-Dioxin and Related Halogenated Aromatic Hydrocarbons: Examination of the Mechanism of Toxicity. *Ann. Rev. Pharmacol. Toxicol.* **1982,** *22,* 517–554.

9. Olujimi, O. O.; Fatoki, O. S.; Odendaal, J. P.; Okonkwo, J. O. Endocrine Disrupting Chemicals (Phenol and Phthalates) in the South African Environment: A Need for More Monitoring. *Water SA.* **2010,** 36, 671–682.

10. Lee, K. T.; Lin, C. H.; Lu, S. Y. SnO_2 Quantum Dots Synthesized with a Carrier Solvent Assisted Interfacial Reaction for Band-Structure Engineering of TiO_2 Photocatalysts. *J. Phys. Chem. C* **2014,** *118,* 14457–14463.

11. Singhal, B.; Porwal, A.; Sharma, A.; Ameta, R.; Ameta, S. C. Photocatalytic Degradation of Cetylpyridinium Chloride Over Titanium Dioxide Powder. *J. Photochem. Photobiol. Chem.* **1997,** *108* (1), 85–88.

12. Sood, S.; Kumar, S.; Umar, A.; Kuar, A.; Mehta, S. K.; Kansal, S. K. TiO_2 Quantum Dots for the Photocatalytic Degradation of Indigo Carmine Dye. *J. Alloys Compd.* **2015,** *650,* 193–198.

13. Kaur, A.; Umar, A.; Kansal, S. K. Sunlight-Driven Photocatalytic Degradation of Non-Steroidal Anti-inflammatory Drug Based on TiO_2 Quantum Dots. *J. Colloid Interf. Sci.* **2015,** *459,* 257–263.

14. Yang, S.; Hou, Y.; Xing, J.; Zhang, B.; Tian, F.; Yang, X. H.; Yang, H. G. Ultrathin SnO_2 Scaffolds for TiO_2-Based Heterojunction Photoanodes in Dye Sensitized Solar Cells: Oriented Charge Transport and Improved Light Scattering. *Chem. Europ. J.* **2013,** *19,* 9366–9370.

15. Tiwana, P.; Docampo, P.; Johnston, M. B.; Snaith, H. J.; Herz, L. M. Electron Mobility and Injection Dynamics in Mesoporous ZnO, SnO_2, and TiO_2 Films Used in Dye-Sensitized Solar Cells. *ACS Nano* **2011,** *5,* 5158–5166.

16. Babu, B.; Cho, M.; Byon, C.; Shim, J. One Pot Synthesis of Ag-SnO_2 Quantum Dots for Highly Enhanced Sunlight-Driven Photocatalytic Activity. *J. Alloys Comp.* **2018,** *731,* 162–171.

CHAPTER 2

Biodiversity as a Source of Drugs: *Cordia, Echinacea, Tabernaemontana,* and *Aloe*

FRANCISCO TORRENS[1*] and GLORIA CASTELLANO[2]

[1]*Institut Universitari de Ciència Molecular, Universitat de València, Edifici d'Instituts de Paterna, P.O. Box 22085, E-46071 València, Spain*

[2]*Departamento de Ciencias Experimentales y Matemáticas, Facultad de Veterinaria y Ciencias Experimentales, Universidad Católica de Valencia San Vicente Mártir, Guillem de Castro-94, E-46001 València, Spain*

[*]*Corresponding author. E-mail: torrens@uv.es*

ABSTRACT

Plants from tropical forest served as a source of medicines for people from the tropics for centuries. Many therapeutic agents were derived from tropical forest species. Brazilian biodiversity was the source of active molecules used by the *pharma*. The combination of a rich traditional medicine and biodiversity places Brazil in a privileged position for drug discovery. Brazilian government created organizations and laws to protect and stimulate biodiversity local exploitation. Aché launched drug discovery via pharmacology and modern chemical, and pharmacological screening to discover active molecules, for example, Acheflan® from *Cordia verbenacea* for inflammatory processes. Preparations of *Echinacea purpurea* were used for centuries for common cold. Research showed that the main active constituents in ethanol preparations, alkylamides, act distinctively on the immune system with a specific binding on the cannabinoid-2 receptor, which leads to an immune modulatory and anti-inflammatory mode of action. Less was known about its direct antiviral properties. It was investigated in vitro the antiviral effects of Echinaforce®, an ethanol tincture from *E. purpurea* herba/radix (95:5), which inhibited effectively H3N2 influenza, and herpes simplex and respiratory syncytial viruses. Experiments on the antibacterial activity versus strains, which play a major role in infections of the upper respiratory tract, revealed good antibacterial effects.

2.1 INTRODUCTION

Plant species from tropical forest served as a source of medicines for people from the tropics for centuries. A large number of modern therapeutic agents presently in use were derived from tropical forest species. For many years, Brazilian biodiversity was the source of active molecules used by pharmaceutical companies. The combination of a rich traditional medicine and biodiversity places Brazil in a privileged position for drug discovery programs. Brazilian government created a series of organizations and laws to protect and stimulate biodiversity local exploitation. Aché laboratories launched a drug discovery research program via pharmacology, and modern chemical and pharmacological screening to discover active molecules. Acheflan® development from *Cordia verbenacea* (Boraginaceae), which is used in folk medicine for the treatment of several inflammatory processes, is an example of the successful story. Essential oil (EO) from *Aniba rosaeodora* var. *amazonica* Ducke syn. *A. duckei* Kostermans, Lauraceae is used in essences industry as the main ingredient of Chanel No. 5 perfume.

Preparations of the eastern purple coneflower *Echinacea purpurea* were used for centuries, particularly for the treatment of common cold. Research showed that the main active constituents in ethanol EtOH preparations, alkylamides, act distinctively on the immune system with a specific binding on the cannabinoid (CB)-2 receptor, which leads to an immune modulatory and anti-inflammatory mode of action. Less was known about the direct antiviral properties of *Echinacea*. It was investigated in vitro the antiviral effects of Echinaforce®, an EtOH tincture from *E. purpurea* herba/radix (95:5), which, even in low concentrations (1:52,000), inhibited effectively influenza type-H3N2, and herpes simplex and respiratory syncytial viruses. Experiments on the antibacterial activity versus strains,

Biodiversity as a Source of Drugs

which play a major role in infections of the upper respiratory tract, revealed good antibacterial effects. Structure-based classification and anti-cancer effects of plant metabolites were reviewed.[1]

Earlier publications in *Nereis*, and so on classified yams,[2] lactic acid bacteria,[3] fruits,[4] food spices,[5] and oil legumes[6] by principal component, cluster, and metaanalyses. The molecular classifications of 33 phenolic compounds derived from the cinnamic and benzoic acids from *Posidonia oceanica*,[7] 74 flavonoids,[8] 66 stilbenoids,[9] 71 triterpenoids and steroids from *Ganoderma*,[10] 17 isoflavonoids from *Dalbergia parviflora*,[11] 31 sesquiterpene lactones (STLs),[12,13] and STL artemisinin derivatives[14] were informed. A tool for interrogation of macromolecular structure was reported.[15] Mucoadhesive polymer hyaluronan favors transdermal penetration absorption of caffeine.[16,17] Polyphenolic phytochemicals in cancer prevention and therapy, bioavailability, and bioefficacy were reviewed.[18] From Asia to Mediterranean, soya bean, Spanish legumes, and commercial *soya bean* principal component, cluster, and meta-analyses were informed.[19] Natural antioxidants from herbs and spices improved the oxidative stability and frying performance of vegetable oils.[20] The relationship between vegetable oil composition and oxidative stability was revealed via a multifactorial approach.[21] It was informed chemical and biological screening approaches to phytopharmaceuticals,[22] cultural interbreeding in indigenous and scientific ethnopharmacology,[23] ethno-botanical studies of medicinal plants, underutilized wild edible plants, food, medicine,[24] phylogenesis by information entropy, avian birds, and 1918 influenza virus.[25] The aim of this work is to review the past, present, and future of biodiversity as a source of new drugs, *Cordia verbenacea* (Acheflan®), triple way of action of eastern purple coneflower *Echinacea purpurea* (Echinaforce®), bioactivities of indole alkaloids isolated from the leaves of *Tabernaemontana elegans*, *Aloe* genus plants, farm and food applications, phytopharmacotherapy, bone health, and natural products (NPs).

2.2   BRAZILIAN BIODIVERSITY AS A SOURCE OF NEW DRUGS: PAST/PRESENT/FUTURE

One of the main molecules used for glaucoma treatment, pilocarpine, is extracted from the leaves of different *Pilocarpus* spp., mainly *P. jaborandi*

Holmes (Rutaceae, cf. Fig. 2.1c).[26] From *Chondrodendron tomentosum* Ruiz *et* Pav. (Menispermaceae), plant used by different tribes of Indians of Amazon basin for preparing a powerful arrow poison (*curare*), D-tubocurarine was discovered, which is an antagonist specific of nicotinic receptor, used to paralyze the muscles of patients under anesthesia (Fig. 2.1a). Nowadays, D-tubocurarine was substituted by other muscle relaxants. Most new compounds, for example, Atracurium®, were inspired in the structure of the compound isolated from plants used since centuries by Amazon Indians. An alkaloid with emetic properties (emetine, Fig. 2.1b) was isolated from the roots of *Psychotria ipecacuanha* (Brot. Stokes) or *Cephaelis ipecacuanha*, Rubiaceae. Emetine is used to induce vomit and presents different therapeutic applications. Since centuries, Guarani Indians of Paraguay and South Brazil used plant *Stevia*, mainly *S. rebaudiana* Bertoni, family Asteraceae, to sweeten their traditional beverage based on mate (*Ilex paraguariensis* A. St. Hil Aquifoliaceae). From *S. rebaudiana*, stevioside was isolated, noncaloric sweetener of great commercial value (Fig. 2.1d). From the venom of Brazilian snake *Bothrops jararaca*, a new class of compounds for hypertension treatment was discovered, the inhibitors of angiotensin-converting enzyme (Fig. 2.1e). The discovery carried out by Sérgio Henrique Ferreira gave rise to two drugs: Captopril® and Enalapril®.

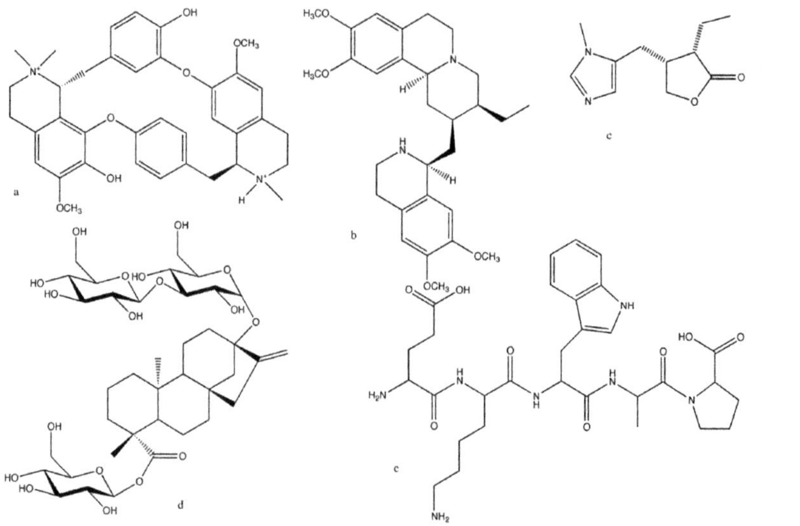

FIGURE 2.1 (a) D-tubocurarine; (b) emetine; (c) pilocarpine; (d) stevioside; (e) BPP5α-peptide from jararaca venom.

Biodiversity as a Source of Drugs

2.3 THE TRIPLE WAY OF ACTION OF *Echinacea purpurea* (ECHINAFORCE®)

The first to show a possible molecular mechanism of action of *Echinacea* was Juerg Gertsch, who discovered that in EtOH preparations of *E. purpurea*, alkylamides (cf. Fig. 2.2) are the components responsible for the immune-modulator effect.[27]

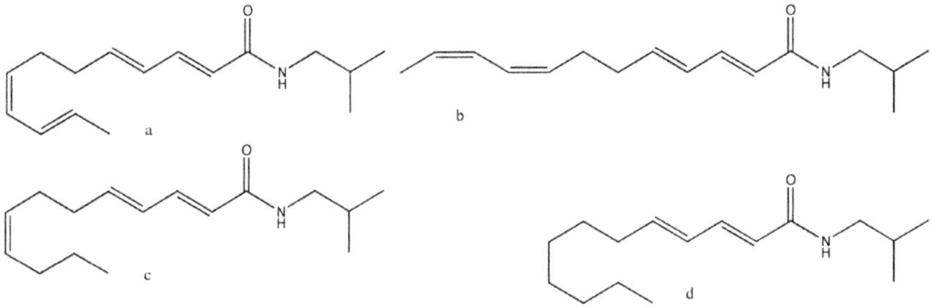

FIGURE 2.2 Isobutylamides of (a) dodeca-2E,4E,8Z,10E-tetraenoic; (b) dodeca-2E,4E,8Z,10Z-tetraenoic; (c) dodeca-2E,4E,8Z-trienoic; (d) dodeca-2E,4E-dienoic acids.

2.4 BIOACTIVITIES OF INDOLE ALKALOIDS ISOLATED FROM THE LEAVES OF *T. elegans*

Three known (*1–3*) and a novel (*4*) monoterpene, and three novel β-carboline indole alkaloids (*5–7*) were isolated from the methanol MeOH extract of *Tabernaemontana elegans* leaves, and their structures were elucidated via spectroscopic experiments (e.g., nuclear magnetic resonance, NMR, mass spectrometry, MS, ultraviolet, UV, infrared, IR, cf. Fig. 2.3).[28] The presence of an $N_4-C_{16}-C_{15}$ linkage in *4* is an unprecedented skeletal feature. The isolated monoterpene and β-carbolines indole alkaloids were studied for their apoptosis induction activity in human hepatocyte (HuH)-7 cancer cells. Methods for apoptosis induction studies included the cell viability assays, nuclear morphology assessments and general cysteinyl-directed aspartate-specific protease (caspase)-3-like activity assays. The monoterpene indole alkaloids, tabernaemontanine (*1*), and vobasine (*3*) showed the most promising apoptosis induction profile in HuH-7. Compounds *5–7* were evaluated for their ability to modulate multidrug resistance (MR) in mouse lymphoma cell lines. Compounds *5* and *7* exhibited a weak activity.

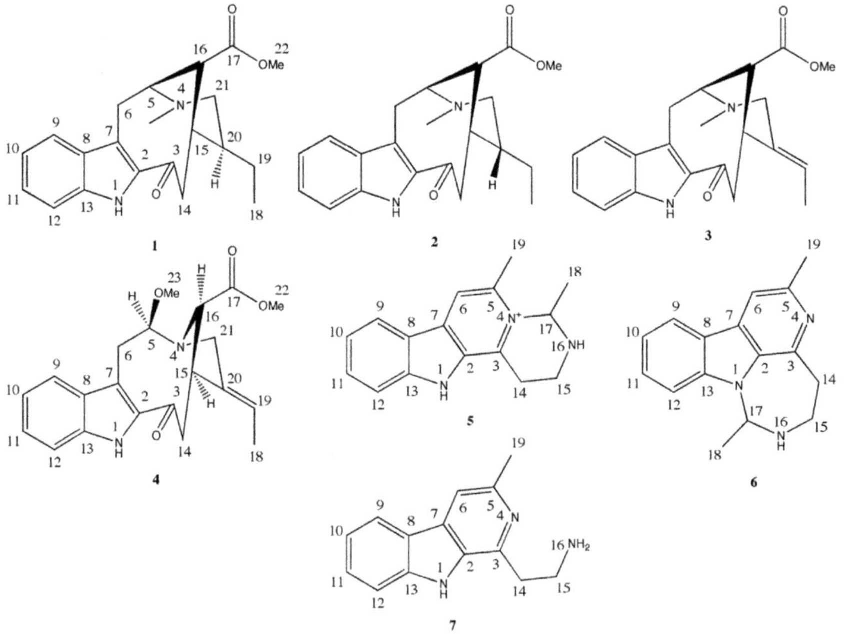

FIGURE 2.3 Monoterpene and β-carboline indole alkaloids isolated from the leaves of *T. elegans*.

2.5 *Aloe* GENUS PLANTS: FARM/FOOD APPLICATIONS/ PHYTOPHARMACOTHERAPY

Aloe vera was reviewed as penetration enhancer.[29] *Aloe* genus plants from farm to food applications and phytopharmacotherapy were revised (cf. Fig. 2.4).[30]

2.6 BONE HEALTH AND NATURAL PRODUCTS: AN INSIGHT

Bone health and NPs were reviewed.[31] Figure 2.5 represents chemical structures of phytochemicals with antiosteoporotic activity.

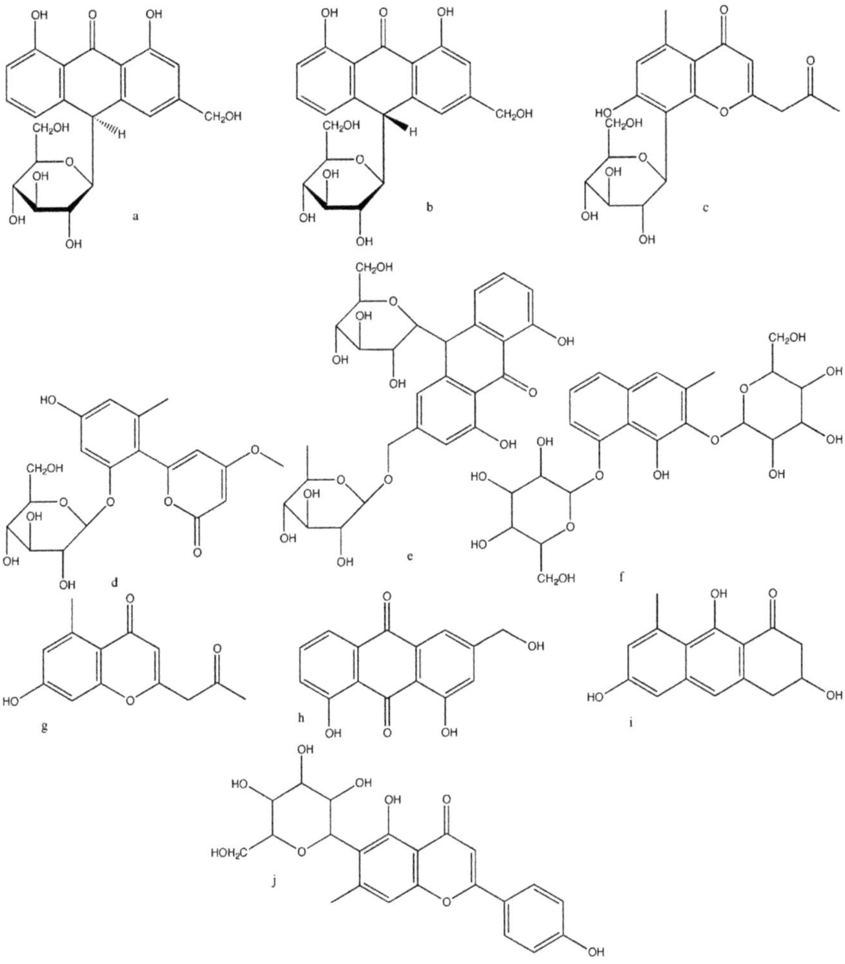

Biodiversity as a Source of Drugs

15

FIGURE 2.4 (a) Aloin A and (b) aloin B; (c) aloesin; (d) aloenin; (e) aloinoside; (f) plicataloside; (g) aloesone; (h) aloe emodin; (i) aloesaponol; (j) isovitexin.

2.7 DISCUSSION

The NPs are a source of active molecules with therapeutic potential. Since centuries, Brazilian biodiversity was source of commercial products explored mainly by foreign companies. After Acheflan® development by Aché laboratories, it was proved that biodiversity can bring specific benefits to developing countries. Acheflan® arrival caused a considerable

16 Green Chemistry and Biodiversity: Principles, Techniques, and Correlations

change of scene of Brazilian pharmaceutical industry. Companies passed to invest in the development of their own products. Specifically, tropical forests turn into source of benefits for society, making clear that biodiversity and sustainable-development protection can guarantee a competitive future for developing countries in present global market.

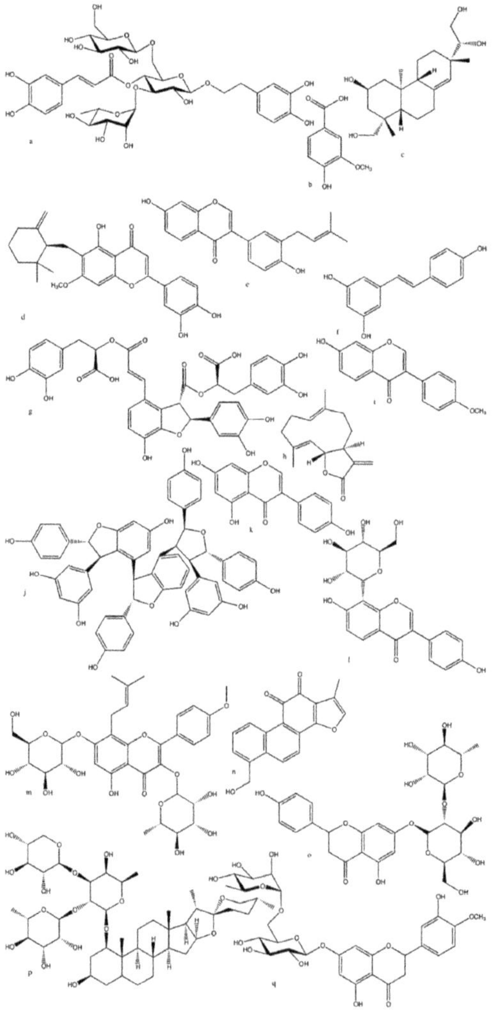

FIGURE 2.5 (a) Echinacoside; (b) vanillic acid; (c) kirenol; (d) ugonin K; (e) neobavaisoflavone; (f) resveratrol; (g) salvianolic acid B; (h) costunolide; (i) formononetin; (j) kobophenol A; (k) genistein; (l) puerarin; (m) icariin; (n) tanshinol; (o) naringin; (p) ophiopogonin D; (q) poncirin.

Biodiversity as a Source of Drugs 17

Vegetal-drugs preparations contain a mixture of components, which confers them the advantage, over their isolated constituents, of presenting greater diversity of activities and manners of action. Anti-inflammatory and immune-modulating, and direct antivitral and antibacterial activities of Echinaforce®, are the base of its clinical effectiveness. Viral infections of the upper respiratory tract, in particular, those caused by influenza, are a rising motive of world worry, especially since viruses from avian or porcine origin transmitted to man.[32] New therapeutic options are searched. A wide agreement exists in the scientific community that treatments for great influenza epidemics or pandemics must be antiviral, anti-inflammatory, and immune-modulating. Taking into account the different manners of action of *E. purpurea*, this can be a treatment option valid not only for common cold but also in infections caused by the influenza virus.

Bone disorders (e.g., osteoporosis, osteoarthrtis) become prevalent with the growth in the aging population. Few conventional therapies exist for the treatment and prevention of osteoporosis (e.g., natural and synthetic estrogens, hormone replacement therapy, Ca^{2+} in combination with vitamin D, bisphosphonates). All the therapies suffer from drawbacks. The adverse effects (e.g., burning sensation, gastrointestinal tract disturbances) limit use. The protective effects by NPs in multifactorial dysmetabolic disease, for example, osteoporosis, could be an effective alternative to overcome side effects of conventional therapy. Osteoporosis could be better treated via a multitarget therapy (e.g., association of multiple herbal drugs and NPs).

2.8 FINAL REMARKS

From the previous results and discussion, the following final remarks can be drawn:

(1) Brazilian biodiversity was source of commercial products explored by foreign companies. After Aché developed Acheflan®, biodiversity can bring specific benefits to developing countries. Acheflan® caused a change of scene of Brazilian pharmaceutical industry. Companies passed to invest in the development of their own products. Tropical forests turn into source of benefits for

society, making clear that biodiversity, and sustainable-development protection can guarantee a competitive future for developing countries in global market.

(2) Vegetal-drugs preparations contain a mixture of components, which confers the advantage of presenting greater diversity of activities and manners of action. Echinaforce® anti-inflammatory/immune-modulating, and direct antivitral/antibacterial activities are base of clinical effectiveness. Viral infections of the upper respiratory tract are a rising motive of world worry. Therapeutic options are searched. A wide agreement exists in the scientific community that treatments for great influenza epidemics/pandemics must be antiviral, anti-inflammatory, and immune-modulating. Taking into account the different manners of action of *E. purpurea*, this can be a treatment option valid not only for common cold but also in infections caused by influenza virus.

(3) Bone disorders become prevalent with the growth in ageing population. Few conventional therapies exist for the treatment and prevention of osteoporosis. All the therapies suffer from drawbacks. The adverse effects limit use. The protective effects of NPs in multifactorial dysmetabolic disease could be an effective alternative to overcome side effects of conventional therapy. Osteoporosis could be better treated via a multitarget therapy.

(4) Medicinal plants were used since ancient times, and are still used as a primary source of medical treatment in developing countries. Plant-derived substances present advantages (e.g., low cost, rapid speed of drug discovery); their main disadvantage is the absence of common international standards for evaluating their quality, efficacy, and safety.

ACKNOWLEDGMENTS

The authors thank support from Generalitat Valenciana (Project No. PROMETEO/2016/094) and Universidad Católica de Valencia *San Vicente Mártir* (Projects No. UCV.PRO.17-18.AIV.03 and 2019-217-001).

KEYWORDS

- antibacterial
- anti-inflammatory
- antiviral
- bone disorder
- immune-modulation
- osteoporosis
- phytochemical

REFERENCES

1. Shin, S. A.; Moon, S. Y.; Kim, W. Y.; Paek, S. M.; Park, H. H.; Lee, C. S. Structure-based Classification and Anticancer Effects of Plant Metabolites. *Int. J. Mol. Sci.* **2018,** *19,* 1–33.
2. Torrens-Zaragozá, F. Molecular Categorization of Yams by Principal Component and Cluster Analyses. *Nereis* **2013,** *2013* (5), 41–51.
3. Torrens-Zaragozá, F. Classification of Lactic Acid Bacteria Against Cytokine Immune Modulation. *Nereis* **2014,** *2014* (6), 27–37.
4. Torrens-Zaragozá, F. Classification of Fruits Proximate and Mineral Content: Principal Component, Cluster, MetaAnalyses. *Nereis* **2015,** *2015* (7), 39–50.
5. Torrens-Zaragozá, F. Classification of Food Spices by Proximate Content: Principal Component, Cluster, Meta-Analyses. *Nereis* **2016,** *2016* (8), 23–33.
6. Torrens, F.; Castellano, G. From Asia to Mediterranean: Soya Bean, Spanish Legumes and Commercial *Soya Bean* Principal Component, Cluster and Meta-Analyses. *J. Nutr. Food Sci.* **2014,** *4* (5), 98–98.
7. Castellano, G.; Tena, J.; Torrens, F. Classification of Polyphenolic Compounds by Chemical Structural Indicators and its Relation to Antioxidant Properties of *Posidonia Oceanica* (L.) Delile. *Match Commun. Math. Comput. Chem.* **2012,** *67,* 231–250.
8. Castellano, G.; González-Santander, J. L.; Lara, A.; Torrens, F. Classification of Flavonoid Compounds by Using Entropy of Information Theory. *Phytochemistry* **2013,** *93,* 182–191.
9. Castellano, G.; Lara, A.; Torrens, F. Classification of Stilbenoid Compounds by Entropy of Artificial Intelligence. *Phytochemistry* **2014,** *97,* 62–69.
10. Castellano, G.; Torrens, F. Information Entropy-based Classification of Triterpenoids and Steroids from *Ganoderma*. *Phytochemistry* **2015,** *116,* 305–313.
11. Castellano, G.; Torrens, F. Quantitative Structure–antioxidant Activity Models of Isoflavonoids: A Theoretical Study. *Int. J. Mol. Sci.* **2015,** *16,* 12891–12906.
12. Castellano, G.; Redondo, L.; Torrens, F. Qsar of Natural Sesquiterpene Lactones as Inhibitors of Myb-Dependent Gene Expression. *Curr. Top. Med. Chem.* **2017,** *17,* 3256–3268.

13. Torrens, F.; Redondo, L.; León, A.; Castellano, G. Structure–activity Relationships of Cytotoxic Lactones as Inhibitors and Mechanisms of Action. *Curr. Drug Discov. Technol.* Submitted for Publication.
14. Torrens, F.; Redondo, L.; Castellano, G. Artemisinin: Tentative Mechanism of Action and Resistance. *Pharmaceuticals* **2017,** *10*, 20.
15. Torrens, F.; Castellano, G. A Tool for Interrogation of Macromolecular Structure. *J. Mater. Sci. Eng. B* **2014,** *4* (2), 55–63.
16. Torrens, F.; Castellano, G. Mucoadhesive Polymer Hyaluronan As Biodegradable Cationic/Zwitterionic-drug Delivery Vehicle. *Admet Dmpk* **2014,** *2*, 235–247.
17. Torrens, F.; Castellano, G. Computational Study of Nanosized Drug Delivery from *Cyclo*dextrins, Crown Ethers and Hyaluronan in Pharmaceutical Formulations. *Curr. Top. Med. Chem.* **2015,** *15*, 1901–1913.
18. Estrela, J. M.; Mena, S.; Obrador, E.; Benlloch, M.; Castellano, G.; Salvador, R.; Dellinger, R. W. Polyphenolic Phytochemicals in Cancer Prevention and Therapy: Bioavailability Versus Bioefficacy. *J. Med. Chem.* **2017,** *60*, 9413–9436.
19. Torrens, F.; Castellano, G. From Asia to Mediterranean: Soya Bean, Spanish Legumes and Commercial *Soya Bean* Principal Component, Cluster and Meta-Analyses. *J. Nutr. Food Sci.* **2014,** *4* (5), 98–98.
20. Redondo-Cuevas, L.; Castellano, G.; Raikos, V. Natural Antioxidants from Herbs and Spices Improve the Oxidative Stability and Frying Performance of Vegetable Oils. *Int. J. Food Sci. Technol.* **2017,** *52*, 2422–2428.
21. Redondo-Cuevas, L.; Castellano, G.; Torrens, F.; Raikos, V. Revealing the Relationship Between Vegetable Oil Composition and Oxidative Stability: A Multifactorial Approach. *J. Food Compos. Anal.* **2018,** *66*, 221–229.
22. Torrens, F.; Castellano, G. Chemical/Biological Screening Approaches to Phytopharmaceuticals. In *Research Methods and Applications in Chemical and Biological Engineering*; Pourhashemi, A., Deka, S. C., Haghi, A. K., Eds.; Apple Academic–Crc: Waretown, NJ, in Press.
23. Torrens, F.; Castellano, G. Cultural Interbreeding in Indigenous/Scientific Ethnopharmacology. In *Research Methods and Applications in Chemical and Biological Engineering*; Pourhashemi, A., Deka, S. C., Haghi, A. K., Eds.; Apple Academic–Crc: Waretown, NJ, in Press.
24. Torrens, F.; Castellano, G. Ethnobotanical Studies of Medicinal Plants: Underutilized Wild Edible Plants, Food And Medicine. In *Innovations in Physical Chemistry*; Haghi, A. K., Ed.; Apple Academic–Crc: Waretown, NJ, in Press.
25. Torrens, F.; Castellano, G. Phylogenesis by Information Entropy: Avian Birds and 1918 Influenza Virus. In *Modelling & Simulation*; Turner, S., Yunus, J., Eds.; IEEE: London, 2008; Vol. 2, pp 1–2.
26. Queiroz, E. F.; Faro, R. D. R. D. A.; Melo, C. A. Brazilian Biodiversity as a Source of New Drugs: Paste, Present and Future. *Rev. Fitoterapia* **2009,** *9* (S1), 31–35.
27. Suter, A. The Triple Way of Action of *Echinacea Purpurea* (Echinaforce®): Antivirial, Antibacterial and Antiinflammatory-Immune-Modulatory Properties. *Rev. Fitoterapia* **2009,** *9* (S1), 63–68.
28. Mansoor, T. A.; Ramalhete, C.; Ramalho, R. M.; Mulhovo, S.; Molnár, J.; Rodrigues, C. P. M.; Ferreira, M. J. U. Biological Activities of Indole Alkaloids Isolated from the Leaves of *Tabernaemontana Elegans*. *Rev. Fitoterapia* **2009,** *9* (S1), 129–129.

29. Sharma, K.; Mittal, A.; Chauhan, N. *Aloe Vera* as Penetration Enhancer. *Int. J. Drug Dev. Res*. **2015**, *7*, 31–43.
30. Salehi, B.; Albayrak, S.; Antolak, H.; Kregiel, D.; Pawlikowska, E.; Sharifi-Rad, M.; Uprety, Y.; Fokou, P. V. T.; Yousef, Z.; Zakaria, Z. A.; Varoni, E. M.; Sharopov, F.; Martins, N.; Iriti, M.; J. Sharifi-Rad, J. *Aloe* Genus Plants: From Farm to Food Applications and Phytopharmacotherapy. *Int. J. Mol. Sci.* **2018**, *19*, 1–49.
31. Suvarna, V.; Sarkar, M.; Chaubey, P.; Khan, T.; Sherje, A.; Patel, K.; Dravyakar, B. Bone Health and Natural Products—An Insight. *Front. Pharmacol.* **2018**, *9*, 1–12.
32. Grisolía, S., Ed. *La Gripe Aviaria: Un Reto De Salud Pública*; Universidad De Castilla–La Mancha: Cuenca, Spain, 2006.

CHAPTER 3

Biodiversity: Loss and Conservation

ANAMIKA SINGH[1] and RAJEEV SINGH[1,2*]

[1]*Department of Botany, Maitreyi College, University of Delhi, Delhi, India*

[2]*Department of Environmental Studies, Satyawati College, University of Delhi, Delhi, India*

Corresponding author. E-mail: 10rsingh@gmail.com

ABSTRACT

Variety and number of plants and animals in an ecosystem is known as biodiversity for specific ecosystem. In an ecosystem, the living organisms interact with each other along with micro-organisms. Biodiversity of ecosystem is mainly due to genetic variations among organisms and these variations are influenced by environmental factors. Species within any ecosystem are linked by food chains and loss of species or biodiversity indicates breaking of the link in the food chain. Biodiversity is actually a natural wealth of the planet and it fulfills all the requirements of any organism. Due to excess utilization of natural resources the loss of biodiversity occurs, so there is a need of conservation and protection of environment.

3.1 INTRODUCTION

Biodiversity is a term refers to variability of life on earth. The variability occurs between species and with species and it also reflects variations in ecosystem.[1] Biodiversity is measurer of different types of organisms present in an ecosystem. Biodiversity is actually variation of gene content of any ecosystem. These variations are due to adaptation of environment and it varies from equator to pole.[2,3] Biodiversity is just because of change in altitude and latitude of earth. Biodiversity is actually distribution of flora and fauna on earth.[4,5] These flora and fauna can be of different variety and number within a population. Strength of biodiversity or the variation in biodiversity depends upon the adaptation of environment. Drastic and rapid change in environment always leads to extinction of species which ultimately effects the environment.[6-8] In terms of biology, biodiversity is the total gene content of an ecosystem and species[9] An advantage of this definition is that it seems to describe most of the circumstances and presents a unified view of the traditional types of biological variety previously identified.

3.2 MEASUREMENT OF BIODIVERSITY

Biodiversity is a multidimensional structure and it is really tough to measure.[10] An area can be defined as rich or poor in biodiversity.

Biodiversity can be measured as:

(1) Number of genes, number of populations, species, or taxa in an area.
(2) Evenness: It is frequency of allele, number, and relative frequency. Along with this genetic diversity is also a measurement of heterozygosity of allele.
(3) Difference: It is difference in alleles within a population. Generally, a population contains similar types of alleles but a bit diversion always creates different level of genetic structure within a population. These differences cause generation of sub species, it has an important evolutionary significant. Biodiversity can be measured by different methods but most important one is species richness.[10,11]

Species are the basic unit of biodiversity. It thus makes biological sense to measure species richness rather than a higher taxonomic grouping. It is often easier to count the number of species compared to other measures of biodiversity.[11]

Biodiversity 25

3.3 ELEMENTS OF BIODIVERSITY

Biodiversity can be viewed under

(1) Genetic diversity: It measures morphological diversity.[12]
(2) Species diversity: It is actually taxonomic diversity.
(3) Community or ecosystem diversity.

Each of this level are composed of three levels: compositional diversity, structural diversity, and functional diversity.

(1) *Compositional diversity*: Number of representative or species present in any level of diversity. Number of species present is a measure of compositional diversity.
(2) *Structural diversity*: It represents sex ratio and age of the individuals of any species present.
(3) *Functional diversity*: Variations in functions performed by the diversity. It measures number of functionally separate species like feeding mechanism, predator, mortality, predator–prey relationship, and so on.

3.3.1 GENETIC DIVERSITY

It is a variation in genetic makeup of individual or population within a species. Within a species genetic diversity can be observed by variation in growth, color, size, resistant to a disease, and so on. Genetic diversity is an important factor which actually maintains the species diversity. It is the total number of genetic characteristics in the genetic makeup of a species. It is a way for a population to adapt changes in concern with environment. Gene has alleles which contain specific characters and these characters adapts to a changing environment. Variation in a population is variation of alleles. With more variation, it is more likely that some individuals in a population will possess variations of alleles that are suited for the environment. Those individuals are more likely to survive to produce offspring bearing that allele. The population will continue for more generations because of the success of these individuals.[13] Genetic and phenetic diversity have been found in all species at the level of protein, DNA, and individual level. Nature is having a nonrandom diversity, heavily structured, and correlated with environmental variation and stress.[14] Genetic and species diversity are interdependent and there is delicate difference between them. Changes in species diversity cause changes in the environment, which leads to force, adaptation to the remaining species. Changes in genetic diversity, mainly loss of species, always lead to a loss of biological diversity.[13] Loss of genetic diversity in species causes extinction of species.[15,16] Genetic diversity is very important for the survival and adaptation of a species.[17] Change in the habitat of a population pressurizes the individual to adapt itself for the survival. It determines the ability to cope with an environmental challenge. Ultimately it selects the species through natural selection.[18] Epidemic and disease cause loss of genetic diversity which effects biodiversity of any area.[19]

3.3.2 SPECIES DIVERSITY

Species diversity is the number of different species by mean of abundance, distribution, function, or interaction. Species richness is number of species present in any area and it also determines the biodiversity of the area. Diversity among species is the measure of biodiversity but there are so many factors to be counted along with this, like relative abundance and area where a particular species is found. If an area is having 10 species and only two are dominating but for biodiversity we cannot ignore rest eight species as they also help to measure the biodiversity of that area.

Species diversity consists of two components:

(1) Species richness: Number of species present in any area.
(2) Species evenness: Abundance or frequency of the species is known as species evenness.[20–22]

3.3.3 COMMUNITY

A community is the different groups of species staying together and interacting with each other. They share common biotic (biological) and abiotic (physical) factors. A community differs by its habitat. A habitat is the set of resources used by a specific species. Community diversity is variation in type, structure, and functions. Ecosystem diversity is a higher level of diversity. Community diversity is aggregation of different species

Biodiversity 27

having separate demands of abiotic factors of any ecosystem. Community diversity largely depends upon abiotic factors like fire, water, soil, topography, and geology. Ecological diversity includes the variation in both terrestrial and aquatic ecosystems. Ecological diversity can also take into account the variation in the complexity of a biological community, including the number of different niches, the number of trophic levels and other ecological processes. An example of ecological diversity on a global scale would be the variation in ecosystems, such as deserts, forests, grasslands, wetlands, and oceans. Ecological diversity is the largest scale of biodiversity, and within each ecosystem, there is a great deal of both species and genetic diversity.[23,24]

3.4 LOSS OF BIODIVERSITY

It is measure of loss of species from any area or ecosystem.

3.4.1 CAUSES FOR LOSS OF BIODIVERSITY

3.4.1.1 HABITAT CHANGE

Humans have the most effected habitat and they occupy and cover most of the areas of earth for their shelter, food, and development. For their demand of food, they cover most of the land for agriculture. Hectares of area of earth have been utilized as agricultural fields which cause loss of habitat for number of plant species. Natural forest, wetlands, estuaries, and mangroves are continually utilized by humans leading to serious loss of shelter for number of species. Wetlands are destroyed due to draining, filling, and pollution. Most of the species utilized the large areas like beer and wild cats as they cover outer areas of forest for food search and interior of forest for breeding. It is known as habitat fragmentation. Loss of areas affects their natural habitat.

3.4.1.2 CLIMATE CHANGE

Recent drastic changes in climate, especially in warmer regions with high temperatures, had significant impacts on biodiversity and ecosystems, which includes changes in species distributions, population sizes, the timing of reproduction or migration events, and an increase in the frequency of pest and disease outbreaks.

3.4.1.3 INVASIVE SPECIES

The spread of invasive alien species has increased because of increased trade and travel. While increasingly there are measures to control some of the pathways of invasive species, for example, through quarantine measures and new rules on the disposal of ballast water in shipping, several pathways are not adequately regulated, particularly with regard to introductions into freshwater systems.

3.4.1.4 OVEREXPLOITATION

In a marine ecosystem, major loss of fish population is due to demand of fishes and over fishing. Demand for fish as food for people and as feed for aquaculture production is increasing, resulting in increased risk of major, long-lasting collapses of regional marine fisheries. About 50% of the world's commercial marine fisheries are fully exploited while 25% are being overexploited. Frequent cutting down of trees in forest causes loss of shelter of small birds.

3.4.1.5 POLLUTION AND GLOBAL WARMING

High population and over demand of natural resources cause pollution which leads to loss of nature. Pollution can be of several types, air, water, noise, etc., and all affect the natural habitat leading to loss of nature. Nowadays, radiation is having high impact on plants and birds. Pollution causes global warning which effects natural life cycle of animals.

3.4.1.6 POACHING

It is illegal to trade wildlife products by killing endangered animals. Although there is an international ban on trade of products of endangered

Biodiversity   29

species, yet smuggling of wildlife items like furs, horns, tusks, specimens, and herbal products worth millions of dollars per year continues. It also causes loss of species from natural habitat.

3.5   CONSERVATION STRATEGIES

These are the measures used to protect biodiversity.

3.5.1   *MAINTAIN INTACT (VIABLE) LANDSCAPES*

The intent of this strategy is to protect and improve the ecological integrity and long-term viability of the more intact (core) landscapes of the region. Within these areas, priority actions would be to repair historic impacts, remove threats, and reinstate ecological processes.

3.5.2   *REVERSE DECLINES*

This strategy aims to reinstate ecosystems that have been differentially lost in locations where it will meaningfully contribute to stemming species' declines and reinstating critical ecological processes (such as pollination). Within these areas, the priority actions are to reinstate open woodland systems and improve the habitat value of shrubby systems.

3.5.3   *RECOVER THREATENED SPECIES AND ECOLOGICAL COMMUNITIES*

The intent of this strategy is to ensure the long-term persistence of species and ecosystems at immediate risk of extinction in the wild. The actions required to implement this work are specific to individual species and ecosystems, but typically focus on increasing distribution and abundance and halting (or ideally reversing) declining trends. The nature of this work is guided by the current amount of knowledge.

3.5.4   *CONTROL EMERGING THREATS*

This strategy aims to address threats to biodiversity before their impacts are fully realized. A couple of more pervasive threats to the region include climate change and new invasive species.

3.5.4.1 PROTECTED AREAS

The protected areas are biogeographical areas where biological diversity along with natural and cultural resources are protected, maintained, and managed through legal and administrative measures. The demarcation of biodiversity in each area is determined on the basis of climatic and physiological conditions.

In these areas, hunting, firewood collection, timber harvesting, and so on are prohibited so that the wild plants and animals can grow and multiply freely without any hindrance. Some protected areas are: Cold desert (Ladakh and Spiti), Hot desert (Thar), Saline Swampy area (Sundarbans and Rann of Kutch), Tropical moist deciduous forest (Western Ghats and north East), and so on. Protected areas include national parks, sanctuaries, and biosphere reserves. There are 37,000 protected areas throughout the world. As per World Conservation Monitoring Centre, India has 581 protected areas, national parks, and sanctuaries.

3.5.4.2 NATIONAL PARKS

These are the small reserves meant for the protection of wild life and their natural habitats. These are maintained by government. The boundaries are well demarcated and circumscribed. The activities like grazing forestry, cultivation, and habitat manipulation are not permitted in these areas.

3.5.4.3 SANCTUARIES

These are the areas where only wild animals (fauna) are present. Activities like harvesting of timbers, collection of forest products, cultivation of lands, and so on are permitted as long as these do not interfere with the project. That is, controlled biotic interference is permitted in sanctuaries, which allows visiting of tourists for recreation. The area under a sanctuary remains in between 0.61 and 7818 km^2.

Biodiversity 31

3.5.4.4 BIOSPHERE RESERVES

Biosphere reserves or natural reserves are multipurpose protected areas with boundaries circumscribed by legislation. The main aim of biosphere reserve is to preserve genetic diversity in representative ecosystems by protecting wild animals, traditional life style of inhabitant, and domesticated plant/animal genetic resources. These are scientifically managed allowing only the tourists to visit.

Some importance of biosphere reserves is as follows:

(1) These help in the restoration of degraded ecosystem.
(2) The main role of these reserves is to preserve genetic resources, species, ecosystems, and habitats without disturbing the habitants.
(3) These maintain cultural, social, and ecologically sustainable economic developments.
(4) These support education and research in various ecological aspects.

3.6 CONCLUSION

Biodiversity is a wide term and it is specific for any particular area and this diversity is due to effect of environment. The development of flora and fauna largely depends upon survival and finally the survival dominates and grows in next generation. Further these species (plants and animals) modifies their surroundings as per their needs.

KEYWORDS

- **biodiversity**
- **community**
- **conservation**
- **genetic diversity**
- **natural selection**

REFERENCES

1. Clark, M. R.; Schlacher, T. A.; Rowden, A. A.; Stocks, K. I.; Consalvey, M. *Science Priorities for Seamounts: Research Links to Conservation and Management. PLoS One* **2012,** *7,* e29232.
2. Tittensor, D. P.; Mora, C.; Jetz, W.; Lotze, H. K.; Ricard, D. ; Berghe, E. V.; Worm, B. Global Patterns and Predictors of Marine Biodiversity Across Taxa. *Nature* **2010,** *466* (7310), 1098–1101.
3. Myers, Norman; Mittermeier, Russell A.; Mittermeier, Cristina G.; da Fonseca, Gustavo A. B.; Kent, Jennifer; Mittermeier, Cristina G.; Da Fonseca, Gustavo A. B.; Kent, Jennifer. Biodiversity Hotspots for Conservation Priorities. *Nature* **2000,** *403* (6772), 853–858.
4. McPeek, Mark A.; Brown, Jonathan M. Clade Age and Not Diversification Rate Explains Species Richness Among Animal Taxa. *Am. Nat.* **2007,** *169* (4), E97–E106.
5. Rabosky, Daniel L. Ecological Limits and Diversification Rate: Alternative Paradigms to Explain the Variation in Species Richness Among Clades and Regions. *Ecol. Lett.* **2009,** *12* (8), 735–743.
6. Charles, C *Biological Processes Associated with Impact Events ESF IMPACT,* 1st ed.; Springer: Berlin, 2006; pp 197–219.
7. Algeo, T. J.; Scheckler, S. E. Terrestrial-Marine Teleconnections in the Devonian: Links Between the Evolution of Land Plants, Weathering Processes, and Marine Anoxic Events. Philos. *Trans. R. Soc. B Biol. Sci.* **1998,** *353* (1365), 113–130.
8. Bond, D. P. G.; Wignall, P. B. The Role of Sea-level Change and Marine Anoxia in the Frasnian–Famennian (Late Devonian) Mass Extinction. *Palaeogeogr. Palaeoclimatol. Palaeoecol.* **2008,** *263* (3–4), 107–118.
9. Davis, M. L. *Introduction to Environment Engineering (Sie),* 4th ed.; Mcgraw-Hill Education Pvt. Ltd.: India, 2011; p 4.
10. Purvis, A.; Hector, A. Getting the Measure of Biodiversity. *Nature* **2000,** *405,* 212–219.
11. Gaston, K. J.; Spicer, J. I. *Biodiversity: An Introduction,* 2nd ed.; Blackwell, 2004, ISBN: 978-1-405-11857-6 February Wiley-Blackwell.
12. Baco, A. R.; Cairns, S. D. *Comparing Molecular Variation to Morphological Species Designations in the Deep-Sea Coral Narella reveals New Insights Into Seamount Coral Ranges. Plos One* **2012,** *7,* E45555.
13. https://web.archive.org/web.
14. *Nevo, Eviatar.* Evolution of Genome-Phenome Diversity Under Environmental Stress. *Proc. Natl. Acad. Sci. USA* **2001,** *98 (11), 6233–6240.*
15. Groom, M. J.; Meffe, G. K.; Carroll, C. R. *Principles of Conservation Biology,* 3rd ed.: Sunderland, 2006. http://www.sinauer.com/groom/.
16. Tisdell, C. *Socioeconomic Causes of Loss of Animal Genetic Diversity: Analysis and Assessment. Ecol. Econ.* **2003,** *45 (3), 365–376.*
17. *Frankham, Richard* Genetics and Extinction. *Biol. Conserv.* **2005,** *126 (2), 131–140.*
18. Nevo, Eviatar Evolution of Genome-Phenome Diversity Under Environmental Stress. *Proc. Natl. Acad. Sci. USA* **2001,** *98* (11), 6233–6240.

Biodiversity

19. Maillard, J. C.; Gonzalez, J. P. Biodiversity and Emerging Diseases. *Ann. N Y Acad. Sci.* **2006,** *1081,* 1–16.
20. Hill, M. O. Diversity and Evenness: A Unifying Notation and its Consequences. *Ecology* **1973,** *54,* 427–432
21. Tuomisto, H. A Diversity of Beta Diversities: Straightening up a Concept Gone Awry. Part 1. Defining Beta Diversity as a Function of Alpha and Gamma Diversity. *Ecography* **2010,** *33,* 2–22.
22. Tuomisto, H. A Consistent Terminology for Quantifying Species Diversity? Yes, it Does Exist. *Oecologia* **2010,** *4,* 853–860.
23. Mcclain, C. R.; Lundsten, L.; Barry, J.; Devogelaere, A. Assemblage Structure, but not Diversity or Density, Change with Depth on a Northeast Pacific Seamount. *Mar. Ecol.* **2010,** *31* (Suppl. S1), 14–25.
24. Williams, A. et al. *Scales of Habitat Heterogeneity and Megabenthos Biodiversity on an Extensive Australian Continental Margin (100–1000 m Depths). Mar. Ecol.* **2010,** *31,* 222–236.

CHAPTER 4

Aegle marmelos: Nature's Gift for Human Beings

HEMA JOSHI[1], RAJEEV SINGH[2], and ANAMIKA SINGH[3*]

[1]*Department of Botany, Hindu Girls Degree College, Sonipat, Haryana, India*

[2]*Department of Environmental Studies, Satyawati College, University of Delhi, Delhi, India*

[3]*Department of Botany, Maitreyi College, University of Delhi, Delhi, India*

[*]*Corresponding author. E-mail: arjumika@gmail.com*

ABSTRACT

Aegle marmelos (Bael) is also known as Bilva in Sanskrit language. In Ayurveda, it is more commonly known by the same name. This herb has great medicinal, spiritual, and religious significance. Its fruits and leaves are considered sacred and used as offerings to the Hindu Gods like Lord Shiva. This is why it is also known as "Shiva druma" or the tree of Shiva in ancient scriptures. Bilva is also known for its great properties of keeping blood sugar levels under control for which it is being explored far and wide. It is also great for digestion and helps to keep the intestines healthy. Different parts of this herb are used for different purposes. Its fruit is relished as a whole as well as for its chilled juice which is very good to beat the heat in hot and scorching summers like those of northern India. The leaves are also a significant contributor to the many beneficial properties that Bilva herb exhibits. The old yellow fruits of this tree turn green after one year. It contains tannic acid, volatile oil, and mucilaginous liquid. Bael plant acts as a "Sink" for chemical pollutants as it absorbs poisonous gases from atmosphere and makes them inert or neutral. It is a member of plant species group known as "Climate purifier" which emits greater percentage of oxygen in sunlight as compared to other plants. The tree is also considered under the category of "fragrant species" whose flower and volatile vapors neutralize bad smell of petrified organic matter and thus save human life from bacterial attack by making them inert and deodorizing the bad odor of air.

4.1 INTRODUCTION

Bael is known in India from prehistoric time and has been mentioned in the ancient system of medicine. It has a great mythological significance also. Plants are an important source of medicine and a large number of drugs presently in use are derived from plants. According to the Ayurveda, plants have various constituents in them which may be used for the treatment of many ailments. Herbs had been used by all cultures as evident from history but India is supposed to have one of the oldest, richest, and most diverse cultural living traditions that are associated with the use of medicinal plants (Narwal et al., 2001; Atal and Kapoor, 1989). The important advantages claimed for medicinal utilization of plants in various ailments are their safety besides being economical, effective, and their ease of availability. Due to these advantages, the medicinal plants have been commonly utilized by the traditional medical practitioners in their daily practices (Tandon et al., 2004; WHO Survey, 1993). Bael is a spinous deciduous and aromatic tree with long, strong, and axillary spines. This tree grows up to 18 m in height and thickness of tree is about 3–4 ft. Leaves are 3–5 foliate, leaflets are ovate and have typical aroma. Flowers are greenish-white in color and sweet-scented. Fruits are large, woody, grayish yellow, 8–15 celled, and have sweet gummy orange colored pulp. Seeds are compressed, oblong, and numerous, found in aromatic pulp. Different plant parts of bael are beneficial in many problems.

(1) Unripe fruits—balance Kapha and Vata doshas.
(2) Ripe fruits—difficult to digest and aggravate all three doshas.
(3) Roots—improve digestion prevent vomiting and balance all three doshas.
(4) Leaves—relieve pain, dyspepsia, gastritis, and abdominal colic pain. It also balances all three doshas.
(5) Stem—good for the heart, effective in rheumatoid arthritis, and improves secretion of digestive enzymes.
(6) Flowers—relieve diarrhea, dysentery, thirst, and vomiting.

4.2 DIFFERENT NAMES OF BAEL

(1) English name: Apple wood, Bengal quince
(2) Hindi name: Bael, bel patra, Vili, Sriphal
(3) Sanskrit name: Bilva, Sriphal, Pootivat, Shaelpatra, Lakshmiputra, Shiveshta
(4) Gujarati name: Bael, Beelee
(5) Marathi name: Bael
(6) Punjabi name: Bael
(7) Telugu name: Bilvayu, Moredu
(8) Arabic name: Safarjale
(9) Urdu name: Bael
(10) Tamil name: Bilubam

4.3 CLASSIFICATION OF BAEL

(1) Kingdom: Plantae
(2) Division: Magnoliophyta
(3) Class: Magnoliopsida (dicotyledons)
(4) Order: Spanidales
(5) Family: Rutaceae
(6) Genus: Aegle
(7) Species: marmelos

FIGURE 4.1 Plant of *Aegle marmelos*.

4.4 BIOCHEMICAL AND BIOLOGICAL ASPECTS

Bael and its other plant parts are a very significant source of traditional medicine (Fig. 4.1). They have been in use in Ayurveda for a very long time. They strengthen the immune system and fight a variety of infections, diseases, and disorders. The presence of various antioxidants, nutrients, phytochemicals, phenolic compounds, and flavonoids in bael makes it very healthy. Many studies have observed that bael works similar to drugs and does not cause any side effects. Let us have a look at nutrient composition of bael (*Aegle marmelos*).

4.4.1 NUTRIENTS IN BAEL

Bael fruit is a good source of many important minerals, which are essential for human health. Studies have reported that bael fruit contains high amount of potassium and low amount of sodium and thus, it is beneficial for individuals with hypertension. Calcium is the second highest mineral present in bael fruit that is necessary for strong bones, teeth, and nerve impulse transmission. Besides this, other minerals present were iron, copper, zinc, and manganese. It is also a decent source of vitamins such as vitamin B1 (thiamine), vitamin B2 (riboflavin), vitamin B3 (niacin), and vitamin C. The presence of so many nutrients makes it extremely healthy for human consumption.

4.4.2 RICH IN ANTIOXIDANTS

Research has shown that antioxidants present in bael inactivate the free radicals or makes them less reactive and thus guards against reactive oxygen species. Health-promoting compounds present in bael include steroids, terpenoids, flavonoids, phenolic compounds, tannins, alkaloids, and saponins. Out of all these, tannins have a very strong free radical scavenging property and act as a primary antioxidant. The antioxidative characteristics of bael fruit are attributed to the presence of phenolic compounds and phytochemicals such as flavonoids. These flavonoids are responsible for radical scavenging activity of bael. In addition to this, the reducing capacity of a compound is an important indicator of its potential antioxidant activity.

Studies have reported that the reducing power of bael extract was high and it further increased with an increase in the concentration of the extract. Superoxide anion is harmful reactive oxygen species that causes severe damage to the cells. It was found that bael extract had a great ability to scavenge superoxide anion and protect the cells against damage. In simple words, bael fruit pulp is a great source of antioxidant and it can successfully combat oxidative stress.

4.4.3 ABUNDANT IN PHYTOCHEMICALS

Properties of beal as medicine are well known in India. Bael is used for the treatment of various diseases and disorders due to the presence of phytochemicals in it. More than 100 phytochemical compounds have been isolated from different parts of bael plant. Some of them include:

(1) Phenols
(2) Flavonoids
(3) Alkaloids
(4) Terpenoids
(5) Tannins
(6) Steroids
(7) Saponins
(8) Cardiac glycosides

These compounds possess pharmacological (medicine-like) and biological effects against various chronic diseases such as:

(1) Gastrointestinal disorders (gut-related)
(2) Type 2 diabetes mellitus
(3) Cancer
(4) High cholesterol levels
(5) Peptic ulcers
(6) Cardiovascular diseases

4.5 TRADITIONAL KNOWLEDGE—SCRIPTURES

The Bael tree has its origin from Eastern Ghats and Central India. It is indigenous to Indian subcontinent and mainly found in tropical and subtropical regions. The tree is also found as a wild tree in lower range of Himalayas up to an elevation of 500 m. Bael is found growing along foothills of Himalayas, Uttar Pradesh, Bihar, Madhya Pradesh, Chhattisgarh, Uttaranchal, Jharkhand, the Deccan Plateau, and along the East Coast.

Hiuen Tsiang, the Chinese Buddhist pilgrim, who came to India in 1629 A.D. Beal is also grown in some Egyptian Gardens and its fruits has been procured and maintained in citrus collection in Florida. In Bangladesh, the tree has been used for fertility control and antiproliferative, and in Sri Lanka, it has been used for its hypoglycaemic activities. Bael fruit was introduced in Europe in 1959. Hindus hold the tree in great venerations. It is one of the most sacred trees of India.

4.6 USES OF BAEL

Bael possesses various pharmacologic properties and plays an important role in the treatment of various diseases. Besides providing a sweet taste, it also protects the body and improves overall health. Here are some amazing therapeutic advantages of bael:

4.6.1 TYPE 2 DIABETES MELLITUS

Since many years, Bael fruit is widely used in Indian Ayurvedic medicine for the treatment of type 2 diabetes mellitus. Treatment with bael fruit extract has significantly shown to reduce high plasma glucose level, urine glucose level, and oxidative stress. Oxidative stress causes damage to the pancreatic beta cell (producer of insulin) and thus, leads to hyperglycemia or high plasma glucose level. Research has shown that bael fruit reduces the load of free radicals and increases the plasma level of antioxidants. A significant elevation was observed in reduced glutathione and vitamin C levels. It is very well known that reduced glutathione and vitamin C are potent antioxidants that scavenge the free radicals and combats oxidative stress. High dosage of bael fruit extract was even more effective than

Aegle marmelos   41

antidiabetic drug. Therefore, it can be said that bael fruit possesses hypo-glycemic properties and it can be included safely in a diabetic diet.

4.6.2   HIGH CHOLESTEROL AND TRIGLYCERIDE LEVELS

Bael fruit has hypolipidemic activity. Studies have shown that it significantly inhibits elevation in serum cholesterol and triglyceride levels. Furthermore, it was found that bael fruit even increased the level of high-density lipoprotein (HDL) cholesterol (good and healthy cholesterol). Such a rise in HDL cholesterol protects the heart and keeps cardiovascular diseases at bay. Bael leaf extract has also shown to have lipid-lowering property. It suppresses the production of cholesterol, increases its excretion, thus lowering the level of low-density lipopro-tein (LDL) cholesterol (bad and unhealthy), and total cholesterol. It also increases the transport of cholesterol from tissues to the liver and this, in turn, slows down the accumulation of lipids in artery wall and reduces the risk of heart diseases. It regulates lipid metabolism and increases the breakdown of fat.

4.6.3   GASTRIC ULCERS

Research has shown that bael fruit is used as herbal drug and it plays a vital role in the treatment of aspirin-induced gastric ulcers. This is possible due to presence of antioxidants in it. It is found that bael fruit extract increases the protective factors and reduces the aggressive factors that cause ulcers. It elevates the activity of antioxidants such as gluta-thione, catalase, and superoxide dismutase in the stomach and duodenum (part of small intestine) tissues. Such an increase in antioxidants makes the inner lining of stomach and duodenum resistant to harmful actions of irritants. During gastric ulcer, the activity of these antioxidants decreases drastically. Therefore, it becomes essential to consume antioxidant-rich fruit such as bael to restore the action of antioxidants. Furthermore, phenolic compounds present in bael also possess anti-ulcer activities by scavenging free radicals. A decrease in oxidative stress increases the supply of blood, nutrients, and oxygen to stomach tissues and thus restores the lost nutrition.

4.6.4 DIARRHEA

Scientific studies have thrown some light upon anti-diarrheal effect of unripe bael fruits. It has been reported that bael fruit plays a protective role against infectious forms of diarrhea. Bael fruit extract reduced the adherence of bacteria and inhibited the activity of rotavirus and giardia (pathogens that cause stomach infection and diarrhea). Furthermore, it also lowers the colonization of bacteria to inner lining of gut and limits the production of toxins. Thus, bael fruit can be used in the treatment of chronic diarrhea.

4.6.5 LIVER INJURY

It is found that bael leaves have excellent liver-protective effect. Liver diseases are usually caused due to overexposure to toxins or due to infections. Research has shown that bael has antifungal, antibacterial, antimicrobial, and antiviral properties that protect the liver against infections and injury. Toxic agents cause damage to the liver cells, which in turn leads to injury and elevates the level of liver enzymes in the bloodstream. Furthermore, formation of reactive oxygen species and oxidative stress plays an important role in the development of alcoholic liver disease and causes damage to the liver. Bael increases the level of antioxidants in the bloodstream such as glutathione and superoxide dismutase and inhibited the action of toxic agents. It further restores the normal structure and function of liver cells. Treatment with bael extract reduces the level of elevated liver enzymes, which is an important indicator of restoring normal liver function.

4.6.6 INFLAMMATORY BOWEL DISEASE

Inflammatory bowel disease is an intestinal inflammatory disorder and herbal remedies play a wide role in the treatment of such disorders. Bael is a gut-friendly fruit and it is very well known for its anti-inflammatory, antibacterial, and antioxidant properties. Research has shown that treatment with bael extract reduces the severity of intestinal inflammation. Such an effect is seen due to the inhibition of inflammatory markers such as IL_1, IL_6, IL_8, and TNF-α. The anti-inflammatory effect may even

Aegle marmelos

be possible due to the presence of phytochemical constituents such as phenolic compounds, steroids, and flavonoids. In addition to this, bael extract also increased the antioxidant activity of superoxide dismutase. Such an increase is attributed to the presence of antioxidants such as carotenoids, vitamin C, thiamine (vitamin B1), riboflavin (vitamin B2), and niacin (vitamin B3). These antioxidants further protected the inner intestinal lining against damage and inflammation. Furthermore, the effect of **bael fruit extract** on inflammatory bowel disease was similar to that of prednisolone (a drug used in the treatment of inflammatory conditions).

4.6.7 DIURETIC ACTIVITY

Diuretic is a drug that increases the output of urine. This drug is usually given to individuals with increased water retention. Research has found that roots and leaves of bael have natural diuretic activity and it increases the passage of urine. Scientific studies have found that treatment with bael leaves and roots enhanced urine volume and increased the excretion of electrolytes such as sodium and potassium. They work by decreasing the re-absorption of sodium and fluid in the body. Studies even found that diuretic activity of bael roots was higher than that of leaves. Such an activity of bael makes it important in the treatment of congestive heart failure, kidney failure, and high blood pressure where water retention is common.

4.6.8 HYPERTENSION

Research has found that bael fruit is effective in lowering the systolic blood pressure. Such an effect is attributed to the presence of potassium that causes widening of the blood vessels and ensures smooth blood flow. Furthermore, it was reported that bael fruit has low sodium content and thus, it keeps constriction of blood vessels at bay. Its high antioxidant content reduces oxidative stress in kidney and blood vessels and thus helps in keeping the blood pressure within normal range. In addition to this, studies have found no side effects when treated with bael fruit.

4.6.9 IMMUNOMODULATORY POTENTIAL

Research has shown that bael fruit extract stimulates the immune mechanisms and strengthens the immune system. Treatment with bael fruit extract stimulates the immune cells and antibodies, which in turn fight against infections. Such an effect keeps infectious diseases and disorders at bay.

4.7 CONCLUSION

Looking upon wide prospects and potential of bael for various purposes, it is worthwhile to cultivate this plant on large scale especially on unproductive and wasteland. This will help in financial upliftment of poor and landless farmer. Plant carries so many natural ingredients that are helpful to cure number of diseases. Furthermore, scientific research is required to explore the maximum potential of Beal plant. Identification of novel phytochemical, its designing and computational modeling along with receptor binding explores novel areas of biochemical research.

KEYWORDS

- *Aegle*
- antioxidant
- bael
- climate purifier plant
- phytochemicals

REFERENCES

Singh, S. Standardization of Processing Technology of Bael (*Aegle marmelos*Correa). Thesis, Doctor of Philosophy in Horticulture, College of Agriculture CCS, HAU, Hisar, 2000, pp 1–3.

Purohit, S. S.; Vyas, S. P. Medicinal Plant Cultivation: a Scientific Approach. In *Aegle marmelos Correa ex Roxb. (Bael)*; Agrobios: Jodhpur, 2004; pp 280–285.

Agarwal, V. S. Rural Economics of Medicinal Plants: Vegetation in the Forest. In *Drug Plants of India*; Kalyani Publishers: New Delhi, 1997; Vol. 1, pp 1, 6, 44, 45, 102, 103, 129, 160.

Sambamurthy, A. V. S. S.; Subryamanyam, N. S. *Fruits and Nuts, A Textbook of Economic Botany*; Wiley Eastern Limited: New Delhi, 1989; Vol. 4, pp 697–698.

Parmar, C.; Kaushal, M. K. Aegle Marmelos Correa. In *Wild Fruits of the Himalayan Region*; Kalyani Publishers: New Delhi, 1982; pp 1–5.

Kala, C. P. Ethnobotany and Ethnoconservation of *Aegle marmelos* (L.) Correa. *Ind. J. Trad. Know.* **2006,** *5* (4), 537–540.

Karunanayake, Eh.; Welihinda, J.; Sirimanne, S. R.; Sinnadoria, H. Oral Hypoglycaemic Activity of Some Medicinal Plant of Sri Lanka. *J. Ethnopharmacol.* **1984,** *11* (2), 223–231.

Tandon, V.; Thayil, S. Saving Medicinal Plant in South India. *Plants Talk* **1995,** *2*, 16–17.

CHAPTER 5

Seed Growth Method for the Synthesis of Metal Nanoparticles

LAVANYA TANDON, DIVYA MANDIAL, RAJPREET KAUR, and POONAM KHULLAR[*]

Department of Chemistry, B.B.K. D.A.V. College for Women, Amritsar 143005, Punjab, India

[*]Corresponding author. E-mail: virgo16sep2005@gmail.com

ABSTRACT

Seed growth method is a versatile method for the controlled synthesis of the anisotropic nanostructures. Such nanostructures hold promising applications in diverse fields. This method involves a variety of parameters, such as nature and concentration of reducing agent, type of capping agent, the role of Ag^+ ion, mode of addition of reducing agent, etc. However, careful control of these parameters can result in desired nanostructures.

5.1 INTRODUCTION

Metal nanoparticles exhibit unique chemical and physical properties that are different from those of bulk state because of quantum size effects that occur only in specific electronic structures.[1–7] There are two kind of approaches: "top down" and "bottom up." Out of these, the top-down approach suffers from limitation with respect to the control of size and shape of particle,as well as further functionalization.[8] However, the bottom-up approach involves the synthesis of gold nanoparticles (AuNPs) either from chemical or biological reduction.[9]

AuNPs have attracted considerable interest and have potential applications in various fields such as drug delivery, catalysis, bioimaging, sensing, photothermal therapy, nanoelectronics, and in the fabrication of photonic and plasmonic devices.[10–20] The chemical reduction process for the synthesis of metal NPs occurs in two steps, that is, the first step involving nucleation and second one their successive growth. When both the above steps are completed in the same process it is called in-situ synthesis while when they occur separately then it is called seed growth process. This chapter accounts for the understanding of various parameters, such as mode of addition, nature of reducing agent, role of halide ions, role of Ag^+ ion, effect of nature of surfactant tail, etc., on the synthesis of AuNPs and the detailed mechanism of the role played by each.

5.2 SEED GROWTH METHOD

A typical procedure for the synthesis of AuNPs involves the following steps. First two steps involve reduction of gold salt using reducing agent such as sodium borohydride, ascorbic acid, citric acid, etc., while the second one involves stabilization of such metal nanoparticles by some suitable capping agents, such as trisodium citrate (Turkevich method), dihydrate or a surfactant, such as cetyltrimethylammonium bromide (CTAB), etc.[21] In Turkevich method, it is the carboxy acetone obtained from oxidation of citrate that actually act as the stabilizing agent rather than the citrate itself. Due to the weak reducing strength of the citric acid, such reactions usually occur at a high temperature.

Alkylthiols of various chain lengths are also used as stabilizing agents for the synthesized Au NPs. Thiols were first used in the two-phase Brust–Schiffrin method published in 1994.[22] It is a very useful method due to the (1) highly stable AuNPs, (2) facile synthesis in ambient conditions, (3) narrow dispersity, (4) easy functionalization of the synthesized AuNPs for further application. The Au NPs are stabilized by relatively strong Au-S bond and hence larger S/Au mole ratio results in the small average core sizes. During such reactions, the H-atom of thiol (SH) is lost due to the oxidative addition of S-H bond onto two gold atoms of the AuNP's

Synthesis of Metal Nanoparticles

surface. Further modification of the Brust–Schiffrin method involves the use of p-mercaptophenol in methanol solution without any phase transfer agent.[23]

It is generally observed that any thiol that is soluble in the same solvent as $HAuCl_4$, such as methanol, ethanol, etc., allows the use of single-phase system of Au NP synthesis. Due to the weakness of the gold citrate bond, the citrate stabilized Au NPs are easily functionalized upon substitution of the citrate ligand by stronger thiolated ligand. Au NPs have been also functionalized with thiolated PEG and such nanoparticles have been used as a contrast agent for in vivo X-ray computed tomography imaging.[24]

Now we will discuss various parameters involved in the seed growth process and how a little change in these parameters change the morphology of the resulting nanoparticles.

5.2.1 EFFECT OF MODE OF ADDITION OF REDUCING AGENT

Jana et al.[25] used 12 nm seeds for the synthesis of AuNPs using ascorbic acid as the reducing agent. Different parameters were studied such as mode of addition (slow/fast) of ascorbic acid seed/metal concentration as well as different types of reducing agents. Mostly spherical particles are produced except for rods in the absence of seed. It was observed that the presence of seeds under these conditions appears to give rise to more nucleation events. When the different modes of addition are compared, it is observed that the sudden or the fast appearance of reducing agents, such as ascorbic acid in the seed/gold salt solution promotes the formation of more seeds instead of growth. On the contrary, slow addition of more dilute concentrations of ascorbic acid should suppress further seed formation and promote growth. For those particles in which ascorbic acid was added slowly, the plasmon band gradually red shifts as the seed concentration decreases, indicating large average particle diameter. When the reducing agent was added sufficiently slowly, no additional nucleation events took place (Fig. 5.1).

FIGURE 5.1 Representative electron micrographs of Au particles formed by slow addition of ascorbic acid.
Source: Reprinted with permission from Ref. [25]. © 2001 American Chemical Society.

While using a strong reducing agent such as $NaBH_4$, there is little evidence for seed-mediated nucleation because being strong reducing agents $NaBH_4$ is capable of producing small AuNPs even at room temperature. However, the presence of 12 nm gold seeds increases the average product particle size and widens the size distribution, indicating the particle growth. Some researchers have observed that the presence of $AgNO_3$ promotes the rod formation with a rod-like surfactant, the reason is not clear.[26] Others found that the addition of $AgNO_3$ to the seeding growth solutions without any rod-like surfactant eliminates the rod-like and plated gold nanoparticles. It is speculated that in the presence of halide counterions (Cl^-, Br^-(CTAB)),[26] AgX salts may form and serve as nucleation centers that complicate the kinetics of AuNP formation to favor spherical gold nanoparticles.

It is observed that the presence of seeds can give product smaller than the seed, suggesting that the seed promotes nucleation. But careful control of regent addition can solve this problem, as slow addition of a dilute solution of weak reducing agent will suppress additional nucleation and larger particles with improved monodispersity in the size range of 20–100 nm are obtained. A two-step mechanism to explain the narrow size distribution of the gold nanoparticles has been proposed. The first step involves nucleation that is a slow process, which is followed by simultaneous burst of nucleation and growth mediated by the early nucleation. Initially, nucleation increases rapidly, it slows down and eventually stops as the growth predominate. Thus particle size can be controlled by varying seed to metal salt, at low concentration of seed, additional nucleation events occur metals that undergo the size distribution and slower addition of reducing agent inhibits additional nucleation but promotes nonspherical products. The use of preformed metallic seeds could lead to further nucleation during the growth part of the reaction and, hence, uniform size cannot be achieved (Fig. 5.2).

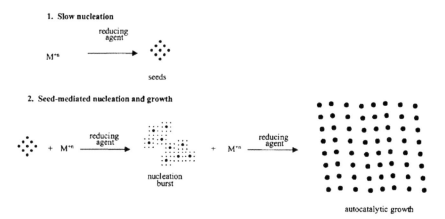

FIGURE 5.2 Generalized two-step mechanism for solution-phase Au nanoparticle synthesis.
Source: Reprinted with permission from Ref. [25]. © 2001 American Chemical Society.

In seed growth method there is always a competition between autocatalytic growth and additional nucleation. The success of earlier seed-mediated growth depends upon the dominance of autocatalytic growth over the additional nucleation. The seed can act as a nucleation center and grow due to the reduction of bulk metal ions at their surfaces. The selection of the reducing agent should always be done for that the metal ion reduction takes place only at the surface of the seed particles rather than creating additional nucleation center. Hence particle size depends on

52 Green Chemistry and Biodiversity: Principles, Techniques, and Correlations

the ratio of seed to metal ions. Another approach to control the particle size involves the use of capping agent that strongly inhibits the growth, therefore, resulting in high monodispersity.

5.2.2 TURKEVICH SYNTHESIS

The gold nanoparticle synthesis by citrate reduction of $HAuCl_4$ is known as Turkevich method/Turkevich synthesis, named after John Turkevich who described the reaction in 1951.[27] The nucleation-diffusional growth model proposed by Turkevich et al. is still the most widely accepted mechanism.[28–29] There are two types of mechanisms, that is, seed-mediated growth mechanism and nucleation-growth mechanism. While the former is based on the seed particles which are made up of hundreds of atoms and are stable in nature, the latter mechanism involves the formation of nuclei that is generally composed of only a few atoms. An "organizer model" was established attributing a decisive role of the formation of the large macromolecules of gold ions and reducing agent with nucleation events being induced by dicarboxyacetone, an oxidation product of trisodium citrate formed during the synthesis. In this method, the synthesis solution appeared to be grey-bluish before changing into purple and the ruby red color, which is characteristic for spherical AuNP in the size range of r = 5–15 nm.[30] Large gold aggregates are not formed at any time of the Turkevich synthesis.

The Turkevich synthesis, which is characterized by seed-mediated growth mechanism, consists of four steps. The first step involves the partial reduction of the gold precursor which results in the formation of the small clusters from the gold monomers. During the second step, small seed particles with size >1.5 nm are formed from the clusters. However, the remaining/leftover gold ions get attracted and attached in the electronic double layer (EDL) of these seed particles as coions. The last two steps involve the growth of the gold monomers exclusively on the top of seed particle's surface and it continues until the whole precursor gets consumed. Afterward, particles are not formed. Also in the process, the number of particles remains constant after a short initial phase.

In Turkevich synthesis, the bluish color of the reaction solution at the early stage is most likely caused by the attachment of gold ions in the EDL of seed particles and the change of their electronic properties.[31] We know that the optical properties of the AuNP aggregates significantly varies with the variation in the charging effects.[30,32] Hence, the total number of particles at the end of the synthesis corresponds to then is determined already at the beginning of the synthesis. The relative particles can be determined using SAXS experiments but a sufficient signal-to-noise ratio is required. The colloidal gold gives the particular adsorption peak at 400 nm which corresponds to the atomic gold.

In Turkevich synthesis, the final size is determined by the total number of particles and that actually corresponds to the number of seed particles, that if more and more of Au^0 is available for seed formation then the resulting nanoparticle will be of smaller size. The second path indicates that the large seed particles are formed from more Au atoms and, hence, at the end of the reaction less number of AuNPs will be formed but their size will be large.

The colloidal stability can be better understood as an interplay of attractive Van der Waals and repulsive electrostatic forces between two particles. This stability is best explained by Dejaguin, Landau, Verwey, Overbeek commonly referred to as the DLVO theory. According to this theory, the sum of these opposing forces, the aggregation barrier depends upon the distance between the two particles. This aggregation behavior increases with the increase of particle size and is better explained as stability curves. As long as the aggregation barrier is lower than the thermal energy, E_{kt}, the two particles can overcome the electrostatic repulsions and can aggregate and, hence, results in particles to grow and vice-versa. Hence the final particle size can be determined using stability curve that depends upon the surface charge and on the other parameter, such as the ionic strength and available thermal energy. Thus a Turkevich synthesis uses a weak reducing agent, monomers are produced relatively fast and also smaller the final particle size more the number of seed particles are formed.

In the gold coordination sphere, four anions are coordinated in a square planar geometry, at strong acidic conditions, all four products are occupied by the Cl^- ions, but as the pH is increased Cl^- ligands get successfully exchanged by OH^- ions. At mild acidic, neutral, and mild alkaline conditions, mixtures of two or more gold complex species are obtained. It is also observed that with an increase of the pH value the concentration of species with higher redox potential decreases.

The $[AuCl_4]^-$ complex exhibits two absorbance bands at ~226 and 313 nm which can be assigned to the $p\pi \rightarrow 5d_x^2 - y^2$ and $p\sigma \rightarrow 5dx^2 - y^2$

ligand–metal transition, respectively.[33–36] The absorption bands decrease and show blue shift with an exchange of Cl^- by OH^- ions. Therefore, the absorbance at 313 nm can be used to monitor the kinetics of the $[AuCl_{y-x}OH_x]$ equilibrium to study the temperature effect.

Trisodium citrate/citric acid (Na_3Cit) has three carboxyl groups that can be protonated/deprotonated leading to an acid–base buffer. At low pH value, a fully protonated form is favored and absorbance at 200 nm can be used to monitor the kinetics of the protonation equilibrium upon addition of H^+ ion.

In seed-growth reaction, the buffer effect of Na_3Cit induces the transformation of the reactive $[AuCl_4]$ species into less reactive $[AuCl_{3-x}(OH)_{1+x}]$ species. Citrate species perform multiple roles that include pH mediator, stabilizer, and reducing agent; the buffer action of citrate shifts the gold complex equilibrium toward the less reactive groups as a result of which seed formation stops. Low concentration of citrate results in the polydispersed gold nanoparticles. High citrate concentration results in the larger sized nanoparticles due to decrease in the colloidal stability of the seed particle. Also, this increase in the particle size results in less number of seed particles. During seed growth mechanism, it is the reduction of $[AuCl_4]^-$ that provides Au^0 monomers, therefore, higher initial concentration of $[AuCl^4]^-$ results in greater monomer and hence more small-sized seed particles are formed. The overall reaction time decreases with increasing gold precursor concentration.

The pH value has a significant effect on the size, polydispersity, and morphology of the final particle. It is observed that under the extreme acidic conditions smaller particles are formed and also the total reaction time gets reduced. More seed particles are formed in more acidic $HAuCl_4$ solutions and with the addition of acid (HCl) or base, the colloidal stability decreases that results in the larger size particles and also the reducing power of Na_3Cit is completely lost.

The seed particle size also depends on the temperature. With increasing temperature, the time frame of the seed particle formation becomes small and also $[AuCl_4]^-$ reduces faster.

Inspired by the Turkevich synthesis, Lee-Meisel used trisodium citrate to prepare silver nanoparticles by the chemical reduction of silver nitrate.[37] However, the resulting nanoparticles are much larger ($r =$

30–100 nm) with a high polydispersity and with different morphologies. Spherical particles can be obtained using a stepwise reduction method.[38] It is due to the fact that although Ag$^+$ ions can exist in different species, no chemical process that transforms a reactive into a less reactive silver precursor species is available that triggers the formation of seed particles. As a result of which large, polydispersed, nonuniform-shaped Ag NPs are obtained.

In the seed growth mechanism, the few formed seed particles are polydispersed but grow into monodispersed AuNPs due to continuous monomer supply during growth steps 3 and 4 that could be described as the size-focusing effect.[39–41]

5.2.3 EFFECT OF CONCENTRATION OF ASCORBIC ACID

Teranishi et al. use two-step mediated method for the synthesis of monodispersed polyhedral gold nanoparticles in which cetyltrimethyl ammonium chloride surfactant is used.[42] It is observed that by keeping all the parameters constant, increasing the concentration of the reductant (ascorbic acid), the morphology of AuNPs get varied from octahedral to truncated octahedral, cuboctahedral, truncated cubic, cubic, and finally trisoctahedral structure. It was observed that when the concentration of ascorbic acid is varied, a large number of gold atoms fed to a single Au seed NP accelerated the growth rate of the {111} crystal plane rather than {100} plane[42] (Figs. 5.3, 5.4).

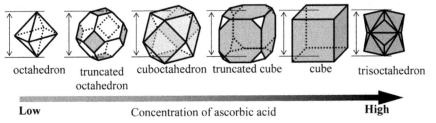

FIGURE 5.3 Schematics of polyhedral NPs. {111}, {100}, and {221} crystal planes are indicated in white, red, and gray, respectively.
Source: Reprinted with permission from Ref. [42]. © 2012 American Chemical Society.

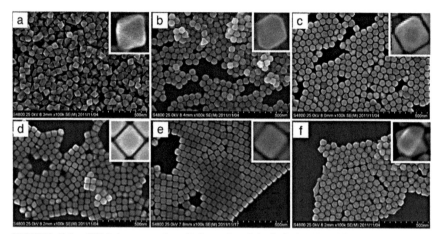

FIGURE 5.4 SEM images of well-faceted polyhedral AuNPs. The shape and the concentration of ascorbic acid for each NP type were (a) octahedron, 0.6 mM; (b) truncated octahedron, 1.2 mM; (c) cuboctahedron, 1.4 mM; (d) truncated cube, 1.8 mM; (e) cube, 2.0 mM; and (f) trisoctahedron, 10 mM. Insets show the enlarged SEM images.
Source: Reprinted with permission from Ref. [42]. © 2012 American Chemical Society.

5.2.4 EFFECT OF CONCENTRATION OF SURFACTANT

Surfactants, such as CTAB, Polyvinyl pyrollidine, play an important role in asymmetric growth of NP leading to nanotriangles. Small changes in the surfactant *cmc* lead to a change in the morphology of the resulting NPs, also the precise control over kinetic reactions especially the reaction rate controlled by the time and sequence of added reagents also changes the morphology of the NPs.[43–46] In the case of Au nanoplates, the presence of two absorption peaks at 709 nm and 1300 nm are observed that corresponds to the inplate dipole and quadrupole plasmon resonance, respectively.

The concentration of surfactant plays a vital role. It is observed that at low concentration of surfactant (CTAB), mainly spherical nanoparticles, with very less yield of Au plate are observed with the increased concentration of 0.025 M triangular gold nanoplates are formed in high yield.[47] Further, an increase in the concentration of up to 0.05 and 0.1 M, hexagonal, circular, and truncated triangular Au plates are observed. Therefore a fine balance of CTAB concentration is must for the high yield synthesis of uniform triangular Au nanoplates. It is observed that

Synthesis of Metal Nanoparticles 57

multiple seed growth in comparison to single step growth offer avoids secondary nucleation.[48–49] Also the short interval decreases the probability of secondary nucleation. [48–49]

5.2.5 ROLE OF Ag⁺ ION

Recently, energy dispersive X-ray spectroscopy have shown the presence of silver in significant amounts on fully grown nanorods, so a special optimized geometry for X-ray detection was required.[50] Many studies have shown that a halide containing a surfactant, such as CTAB is crucial for gold nanorod growth, but there is substantial empirical proof that Ag^+ ion is required for the formation of single crystal nanorods.[51–52] It is observed that silver is essential to break the symmetry of the embryonic nanocrystals, which is followed by the formation of single crystal AuNRs.[53] The small size NPs at the embryonic stage are unstable in nature as compared with fully grown nanorods under prolonged exposure to an intense electron beam required for chemical mapping.

A conventional cuboctahedral structure is a single crystal bound by eight {111} and six {100} facets arranged symmetrically that clearly shows that there is no obvious reason for asymmetric growth.[29] However, if the growth is bound on {100} surfaces but allowed on the {111} facets, then the resulting morphology would be cube bound by {100} surfaces. The cuboctahedron as produced by the Wulf model is minimum energy structure for a face-centered cubic metal such as gold.[54–55] In the absence of the surfactant, for gold particles having size range <10 nm is multiply twinned for which the preferred inclusion of lower energy {111} surface facets results in icosahedra and decahedra structures.[56–59] But in the presence of the surfactants, modification of surface energies relative to the conventional thermodynamic consideration occurs and, therefore, results may vary (Fig. 5.4).

In the absence of Ag^+ ion, no second peak at a higher wavelength in the visible spectra is obtained. But in the presence of Ag^+ another peak at the higher wavelength corresponding to the longitudinal surface plasmon resonance, that is, LSPR is observed that is clearly indicative of shape anisotropy (Fig. 5.5).

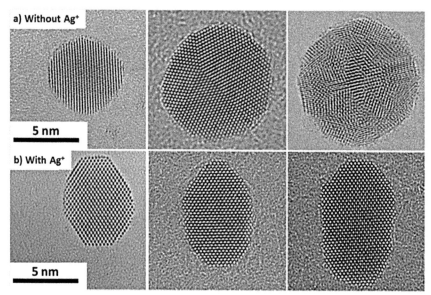

FIGURE 5.5 (a) Gold seed particles overgrown in the absence of Ag^+ remain spherical and may become twinned or multiply twinned. (b) Au seeds overgrown in the presence of Ag^+ (63.4 μM of $AgNO_3$) are observed to become anisotropic.
Source: Reprinted with permission from Ref. [53]. © 2014 American Chemical Society.

The presence of Ag^+ ion results in the stabilization of the higher index truncations that are able to grow into larger facets. The final concentration of nanorod depends upon the concentration of Ag^+ used. These results led to the following three observations:

I. In the presence of Ag^+ ion, the anisotropic growth starts.
II. Symmetry breaking is observed only for the small single crystal particles having a diameter in the range of 4–6 nm.
III. Small asymmetric truncations are observed to appear on the seed particles.

Following steps are proposed to explain these observations:

I Synthesis of seed particles in the presence of cationic surfactant such as CTAB results in single crystal with cuboctahedral morphology that is bound by symmetrically arranged {100} and {111} surfaces (Fig. 5.6).

Synthesis of Metal Nanoparticles

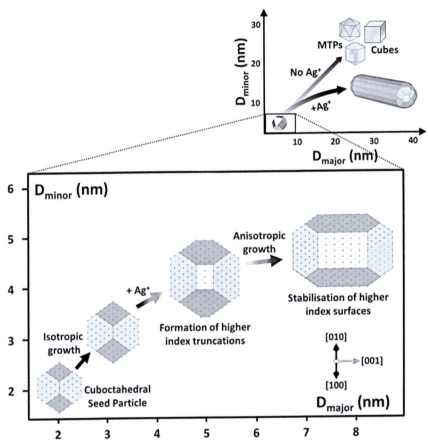

FIGURE 5.6 Schematic representation of the proposed key steps of the symmetry breaking process in single-crystal gold nanorod growth.
Source: Reprinted with permission from Ref. [53]. © 2014 American Chemical Society.

II In the presence of Ag$^+$ ion there occurs initially isotropic growth of seed particles until they touch size diameter in the range of 4–6 nm.

III Till now, small truncating surfaces consisting of few atoms across form nonuniformity at the intersection of {111} facets and these are the potentially preferred sites for silver underpotential deposition.

IV The deposition of Au atoms leads to the stabilization of the higher index truncations. However, growth continues on the lower index surfaces that leads to the truncations, which become side facets in the growing embryonic nanorod structure.

These observations suggest that it is {111} surface sites on which deposition of Au atoms occurs and continued growth on such sites results in the dumbbell-like morphology. The presence of Ag^+ ion also passivates the lower index {111} surfaces hence results in the reduction in the number of single crystal structures. Thus Ag^+ ion plays an important role by stabilizing the truncations that leads to breaking of symmetry, that is, asymmetry.

5.3 FUTURE PERSPECTIVES

This chapter described the role of various important parameters, such as mode of addition of reducing agent, effect of nature, and concentration of reducing agent, Ag^+ ion, etc., in seed growth method. Hence, a little variation in either of the parameter results in different nanostructures. However, with the precise control of such parameters desired nanostructures having potential applications in various fields can be obtained.

KEYWORDS

- **anisotropic nanostructures**
- **capping agent**
- **concentration**
- **reducing agent**
- **seed growth**

REFERENCES

1. Rotello,V. Nanoparticles. *Building Block for Nanotechnology*. Kluwer Academic Publishers: New York, 2004.
2. Schmid, G. *Nanoparticles: From Theory to Application*; Wiley-VCH: Weinheim, 2006.

Synthesis of Metal Nanoparticles

3. Caruso, F. *Colloids and Colloid Assemblies*; Wiley-VCH; Weinheim, 2004.
4. Liz-Marzan, L. M.; Kamat, P. V. *Nanoscale Materials*; Kluwer Academic Publishers: New York, 2003.
5. Burda, C.; Chen, X.; Narayanan, R.; El-Sayed, M. A. Chemistry and Properties of Nanocrystals of Different Shapes. *Chem. Rev.* **2005**, *105*, 1025–1102.
6. Chen, X.; Mao, S. S. Titanium Dioxide Nanomaterials: Synthesis, Properties, Modifications, and Applications. *Chem. Rev.* **2007**, *107*, 2891–2959.
7. Xia, Y., Xiong, Y., Lim, B., Skrabalak, S. E. Shape-controlled Synthesis of Metal Nanocrystals: Simple Chemistry Meets Complex Physics? *Angew. Chem. Int. Ed.* **2009**, *48*, 60–103.
8. Nguyen, D. T.; Kim, D. J.; Kim, K. S. Controlled Synthesis and Biomolecular Probe Application of Gold Nanoparticles. *Micron.* **2011**, *42*, 207–227.
9. Parab, H.; Jung, C.; Woo, M. A; Park, H. G. An Anisotropic Snowflake-like Structural Assembly of Polymer-capped Gold Nanoparticles. *J. Nanopart. Res.* **2011**, *13*, 2173–2180.
10. Eustis S.; El-Sayed M. A. Why Gold Nanoparticles are More Precious than Pretty Gold: Noble Metal Surface Plasmon Resonance and its Enhancement of the Radiative and Nonradiative Properties of Nanocrystals of Different Shapes. *Chem. Soc. Rev.* **2006**, *35*, 209–217.
11. Jain P. K.; Huang X.; El-Sayed I. H.; El-Sayed M. A. Noble Metals on the Nanoscale: Optical and Photothermal Properties and Some Applications in Imaging, Sensing, Biology, and Medicine. *Acc. Chem. Res.* **2008**, *41*, 1578–1586.
12. Murphy, C. J.; Gole, A. M.; Stone, J. W.; Sisco, P. N.; Alkilany, A. M.; Goldsmith, E. C.; Baxter, S. C. Gold Nanoparticles in Biology: Beyond Toxicity to Cellular Imaging. *Acc. Chem. Res.* **2008**, *41*, 1721–1730.
13. Hu, M.; Chen, J.; Li, Z. -Y.; Au, L.; Hartland, G. V.; Li, X.; Marquez, M.; Xia, Y. Gold Nanostructures: Engineering their Plasmonic Properties for Biomedical Applications. *Chem. Soc. Rev.* **2006**, *35*, 1084–1094.
14. Cobley, C. M.; Chen, J.; Cho, E. C.; Wang, L. V.; Xia, Y. Gold Nanostructures: A Class of Multifunctional Materials for Biomedical Applications. *Chem. Soc. Rev.* **2011**, *40*, 44–56.
15. Sardar, R., Funston, A. M., Mulvaney, P. Murray, R. W. Gold Nanoparticles: Past, Present, and Future. *Langmuir* **2009**, *25*, 13840–13851.
16. Ghosh, S. K.; Pal. T., Interparticle Coupling Effect on the Surface Plasmon Resonance of Gold Nanoparticles: From Theory to Applications. *Chem. Rev.* **2007**, *107*, 4797–4862.
17. Hashmi, A. S.; Hutchings, G. J. Gold Catalysis. *Angew. Chem. Int. Ed.* **2006**, *45*, 7896–7936.
18. Sperling, R. A.; Gil, P. R.; Zhang, F.; Zanella, M.; Parak W. J. Biological Applications of Gold Nanoparticles. *Chem. Soc. Rev.* **2008**, *37*, 1896–1908.
19. Boisselier, E.; Astruc, D. Gold Nanoparticles in Nanomedicine: Preparations, Imaging, Diagnostics, Therapies and Toxicity. *Chem. Soc. Rev.* **2009**, *38*, 1759–82.
20. Giljohann, D. A.; Seferos, D. S.; Daniel, W. L.; Massich, M. D.; Patel, P. C.; Mirkin, C. A. Gold Nanoparticles for Biology and Medicine. *Angew. Chem. Int. Ed.* **2010**, *49*, 3280–3294.

21. Turkevich, J., Stevenson, P. C., Hillier, J. A. Study of the Nucleation and Growth Processes in the Synthesis of Colloidal Gold. *Discuss. Faraday Soc.* **1951**, *11*, 55–75.
22. Brust, M.; Walker, M.; Bethell, D.; Schiffrin, D. J.; Whyman, R. J.; Synthesis of Thiol-derivatised Gold Nanoparticles in a Two-phase Liquid–Liquid system. *J. Chem. Soc. Chem. Commun.* **1994**, *7*, 801–802.
23. Brust, M.; Fink, J.; Bethell, D.; Schiffrin, D. J., Kiely, C. J. Synthesis and Reactions of Functionalised Gold Nanoparticles. *J. Chem. Soc. Chem. Commun.* **1995**, *23*, 1655–1656.
24. Kim, D. K.; Park, S. J.; Lee, J. H.; Jeong, Y. Y.; Jon. S. Y.; Antibiofouling Polymer-coated Gold Nanoparticles as a Contrast Agent for In Vivo X-Ray Computed Tomography Imaging. *J. Am. Chem. Soc.* **2007**, *129*, 7661–7665.
25. Jana, N. R.; Gearheart, L.; Murphy, C. J. Evidence for Seed-Mediated Nucleation in the Chemical Reduction of Gold Salts to Gold Nanoparticles. *Chem. Mater.* **2001**, *13*, 2313–2322.
26. Yu, Y. Y.; Chang, S. S.; Lee, C. L.; Wang, C. R. C. Gold Nanorods: Electrochemical Synthesis and Optical Properties. *J. Phys. Chem. B* **1997**, *101*, 6661–6664.
27. Turkevich, J.; Stevenson, P. C.; Hillier, J. A Study of the Nucleation and Growth Processes in the Synthesis of Colloidal Gold. *Discuss. Faraday Soc.* **1951**, *11*, 55–75.
28. Kumar, S.; Gandhi, K. S.; Kumar, R. Modeling of Formation of Gold Nanoparticles by Citrate Method. *Ind. Eng. Chem. Res.* **2007**, *46*, 3128–3136.
29. Patungwasa, W.; Hodak, J. H. pH Tunable Morphology of the Gold Nanoparticles Produced by Citrate Reduction. *Mater. Chem. Phys.* **2008**, *108*, 45–54.
30. Hendel, T.; Wuithschick, M.; Kettemann, F.; Birnbaum, A.; Rademann, K.; Polte, J. In Situ Determination of Colloidal Gold Concentrations with UVV is Spectroscopy: Limitations and Perspectives. *Anal. Chem.* **2014**, *86*, 11115–11124.
31. Polte, J. Fundamental Growth Principles of Colloidal Metal Nanoparticles - A New Perspective. *Cryst Eng Comm.* **2015**, *17*, 6809–6830.
32. Daniel, M. C.; Astruc, D. Gold Nanoparticles: Assembly, Supramolecular Chemistry, Quantum-Size-Related Properties, and Applications toward Biology, Catalysis, and Nanotechnology. *Chem. Rev.* **2004**, *104*, 293–346.
33. Ji, X.; Song, X.; Li, J.; Bai, Y.; Yang, W.; Peng, X. Size Control of Gold Nanocrystals in Citrate Reduction: The Third Role of Citrate. *J. Am. Chem. Soc.* **2007**, *129*, 13939–13948.
34. Wang, S.; Qian, K.; Bi, X. Z.; Huang, W. Influence of Speciation of Aqueous $HAuCl_4$ on the Synthesis, Structure, and Property of Au Colloids. *J. Phys. Chem. C.* **2009**, *113*, 6505–6510.
35. Peck, J. A.; Tait, C. D.; Swanson, B. I.; Brown, G. E., Jr. Speciation of Aqueous Gold(III) Chlorides from Ultraviolet/ Visible Absorption and Raman/Resonance Raman Spectroscopies. *Geochim. Cosmochim. Acta* **1991**, *55*, 671– 676.
36. Goia, D. V.; Matijevi, E. Tailoring the Particle Size of Monodispersed Colloidal Gold. *Colloids Surf.* **1999**, *146*, 139–152.
37. Lee, P. C.; Meisel, D. Adsorption and Surface-Enhanced Raman of Dyes on Silver and Gold Sols. *J. Phys. Chem.* **1982**, *86*, 3391–3395.
38. Dong, X.; Ji, X.; Wu, H.; Zhao, L.; Li, J.; Yang, W. Shape Control of Silver Nanoparticles by Stepwise Citrate Reduction. *J. Phys. Chem. C.* **2009**, *113*, 6573–6576.

39. Polte, J.; Ahner, T. T.; Delissen, F.; Sokolov, S.; Emmerling, F.; Thünemann, A. F.; Kraehnert, R. Mechanism of Gold Nanoparticle Formation in the Classical Citrate Synthesis Method Derived from Coupled In Situ XANES and SAXS Evaluation. *J. Am. Chem. Soc.* **2010,** *132,* 1296–1301.
40. Polte, J.; Herder, M.; Erler, R.; Rolf, S.; Fischer, A.; Würth, C.; Thünemann, A. F.; Kraehnert, R.; Emmerling, F. Mechanistic Insights into Seeded Growth Processes of Gold Nanoparticles. *Nanoscale.* **2010,** *2,* 2463–2469.
41. Wuithschick, M.; Birnbaum, A.; Witte, S.; Sztucki, M.; Vainio, U.; Pinna, N.; Rademann ,K.; Emmerling, F.; Kraehnert, R.; Polte. J. Turkevich in New Robes: Key Questions Answered for the Most Common Gold Nanoparticle Synthesis. *ACS Nano.* **2015,** *9,* 7052–7071.
42. Eguchi, M.; Mitsui, D.; Wu, H.; Sato, R.; Teranishi, T. Simple Reductant Concentration-Dependent Shape Control of Polyhedral Gold Nanoparticles and Their Plasmonic Properties. *Langmuir* **2012,** *28,* 9021–9026.
43. Sau, T. K.; Murphy, C. J.; Room Temperature, High-Yield Synthesis of Multiple Shapes of Gold Nanoparticles in Aqueous Solution. *J. Am. Chem. Soc.* **2004,** *126,* 8648–8649.
44. Xie, S., Lu, N., Xie, Z., Wang, J., Kim, M. J., Xia, Y. Synthesis of Pd-Rh Core-Frame Concave Nanocubes and Their Conversion to Rh Cubic Nanoframes by Selective Etching of the Pd Cores. *Angew. Chem. Int. Ed.* **2012,** 51, 10266–10270.
45. Li, J.; Zheng, Y.; Zeng, J.; Xia, Y. Controlling the Size and Morphology of Au@Pd Core–Shell Nanocrystals by Manipulating the Kinetics of Seeded Growth. *Chem.– Eur. J.* **2012,** *18,* 8150–8156.
46. Zhu, C., Zeng, J., Tao, J., Johnson, M. C., Schmidt-Krey, I., Blubaugh, L., Zhu, Y.,Gu. Z., Xia, Y. Kinetically Controlled Overgrowth of Ag or Au on Pd Nanocrystal Seeds: From Hybrid Dimers to Nonconcentric and Concentric Bimetallic Nanocrystals. *J. Am. Chem. Soc.* **2012,** *134,* 15822–15831.
47. Huang, Y.; Ferhan, A. R.; Gao, Y.; Dandapat, A.; Kim, D. H. High-yield Synthesis of Triangular Gold Nanoplates with Improved Shape Uniformity, Tunable Edge Length and Thickness. *Nanoscale.* **2014,** *6,* 6496–6500.
48. Carrot, G.; Valmalette, J. C.; Plummer, C. J. G.; Scholz, S. M.; Dutta, J.; Hofmann, H.; Hilborn, J. G. Gold Nanoparticle Synthesis in Graft Copolymer Micelles. *Colloid. Polym. Sci.* **1998,** *276,* 853–859.
49. Bakshi, M. S., Sachar, S., Kaur, G., Bhandari, P., Biesinger, M. C., Possmayer, F., Petersen, N. O. *Dependence of Crystal Growth of Gold Nanoparticles on the Capping Behavior of Surfactant at Ambient Conditions. Cryst. Growth Des.* **2008,** *8,* 1713–1719.
50. Pérez-Juste, J.; Pastoriza-Santos, I.; Liz-Marzán, L. M.; Mulvaney, P. Gold Nanorods: Synthesis, Characterization and Applications. *Coord. Chem. Rev.* **2005,** *249,* 1870–1901.
51. Nikoobakht, B.; El-Sayed, M. A. Preparation and Growth Mechanism of Gold Nanorods (NRs) Using Seed-Mediated Growth Method. *Chem. Mater.* **2003,** *15,* 1957–1962.
52. Walsh, M. J., Barrow, S. J., Tong, W., Funston, A. M., Etheridge, J. Symmetry Breaking and Silver in Gold Nanorod Growth. *ACS Nano.* **2015,** *9,* 715–724.

53. Jackson, R.; McBride, J. R.; Rosenthal, S. J.; Wright, D. W. Where's the Silver? Imaging Trace Silver Coverage on the Surface of Gold Nanorods. *J. Am. Chem. Soc.* **2014,** *136,* 5261–5263.

54. Henry, C. R. Morphology of Supported Nanoparticles. *Prog. Surf. Sci.* **2005,** *80,* 92–116.

55. Wulff, G. Z. Zur Frage der Geschwindigkeit des Wachsthums und der Auflösung der Krystallflächen. *Kristallogr. Mineral.* **1901,** *34,* 449.

56. Marks, L. D. Surface Structure and Energetics of Multiply Twinned Particles. *Philos. Mag. A.* **1984,** *49,* 81–93.

57. Barnard, A. S.; Young, N. P.; Kirkland, A. I.; van Huis, M. A.; Xu, H. F. Nanogold: A Quantitative Phase Map. *ACS Nano.* **2009,** *3,* 1431–1436.

58. Howie, A.; Marks, L. D. Elastic Strains and the Energy- Balance for Multiply Twinned Particles. *Philos. Mag. A.* **1984,** *49,* 95–109.

59. Walsh, M. J.; Yoshida, K.; Kuwabara, A.; Pay, M. L.; Gai, P. L.; Boyes, E. D. On the Structural Origin of the Catalytic Properties of Inherently Strained Ultrasmall Decahedral Gold Nanoparticles. *Nano Lett.* **2012,** *12,* 2027–2031.

CHAPTER 6

Soil-Protecting Functions of Medicinal Plants: Meadow and Field Weeds

RAFAIL A. AFANAS'EV, GENRIETTA E. MERZLAYA, and
MICHAIL O. SMIRNOV[*]

*Pryanishnikov All-Russian Scientific Research Institute of
Agrochemistry, d. 31A, Pryanishnikova Street, Moscow 127550, Russia*

[*]Corresponding author. E-mail: User53530@yandex.ru

ABSTRACT

As is known, in medicine and veterinary medicine, a large number of natural therapeutic agents and preparations are used, the source for which is an extensive group of plants of more than 20,000 species, including herbaceous forms.[1] Biologically active substances contained in plant biomass, like alkaloids, glycosides, flavonoids, organic acids, saponins, essential oils, gums, and other ingredients, have long been used as medicinal drugs for people and animals. These substances produced by plants serve to protect them from microorganisms, insects, and herbivores. Along with this, it should be noted that the protection of plants from herbivores is limited not only by biochemical substances, but also by a number of other defense mechanisms, their morphophysiological state, including roughness, the presence of spines, pungent odor, etc. Absence of this comprehensive protection would lead not only to the grassland vegetation disappearance but also to soil losses due erosion processes, that is, to the disappearance of the fundamental principle of the existence of the terrestrial animal and plant world. Considering the interaction of soil–plant–animal from a systemic point of view, we can distinguish three main ways to protect the soil from erosion. The authors of this study believe that general protection measures include decrease of natural pasture productivity due to the vegetational change, the manifestation of forage herbs toxicity under adverse growth conditions and at destruction of the sod due to ungulates the appearance of weeds which have poisonous and repellent properties and so as a rule not consumed by the animals; then there will the normal (herbage) restoration. At the same time, it is weeds that in the majority can relate to medicinal plants due to the content of the above-listed biologically active, including poisonous, substances in their composition. The use of such plants for prophylactic or therapeutic purposes by humans and animals is due to the fact that in minor concentrations biologically active substances, according to the Arndt–Schulz law, enhance biological phenomena in organisms and improve their condition. However, with an increase in doses of taken medicinal plants or their medicinal ingredients, first deceleration of life processes, and then oppression, or even death, can occur.[2] It is in natural ecosystems that largely determines the regulatory role of weeds, from the point of view of agriculture, or medicinal plants, from a medical point of view.

6.1 INTRODUCTION

Using a system approach to assess the ecological role of medicinal plants the authors give a new interpretation of the known facts about grassland vegetation reaction to overload of pastures by animals. The reduction in pasture bioproductivity due to the vegetational change, toxicogenic defense reactions against eating of different plant species under adverse weather conditions and abundant appearance of inedible weeds in localities with destroyed sod are evolutionary developed return reactions of herbaceous ecosystems to the demolition. The reduction in pasture bioproductivity due to the vegetational change, toxicogenic defense reactions against eating of different plant species under adverse weather conditions, and abundant appearance of inedible weeds in localities where with destroyed sod are evolutionary developed return reactions of herbaceous ecosystems to the demolition. From these positions we ought to consider mass appearance of weeds on cultivated land or while drastic meadow improving in the case when these weeds are used in folk and official medicine as a source of medicines. Herbaceous ecosystems perceive the violation of natural turf as impact of herbivores animals and try restoring status quo using their funds developed by the evolution. First, it is a "guard" of plants,

Soil-Protecting Functions of Medicinal Plants

which in agriculture are called meadow and field weeds. To combat them in the meadow and in crop sowing, various methods are being developed, including chemical, physical, and biological, since the yield of crops is reduced in the weedy farmlands and the quality of the products is deteriorating. But from an environmental point of view, in natural conditions and, in part, in agriculture, weeds play a completely opposite role, since they preserve the integrity of the soil from erosion processes or its fertility when grazing ungulates excessively affect the phytocenoses.

6.2 MATERIALS AND METHODOLOGY

Materials of investigation were published messages of geobotany, phytosociology, ecology, mead land farming, and other sciences associated with the study of vital activity of plant communities and the results of own monitoring the state and dynamics of meadow phytocenoses.

The methodological principles of generalization of the materials were the system approach[3,4] and the actualism that is widely used in geology.[5] The system approach is to consider of a plant community and his biotope as a functional unit having the properties of self-regulation, protection from external actions through the development of appropriate responses. The method of the actualism is to recreate the history of development of the ecosystems based on learning modern processes and conditions of their functioning as functional properties of plant communities which have been developed in the process of long evolution. This fact is the basis for the retrospective assessment of the conditions and nature of these properties' formation.

6.3 RESULTS AND DISCUSSION

According to modern data system is a dynamic organization of living and nonliving elements in which there are external and internal circulations of substances determining its functional stability.[6] The most important condition for the stability of any system of the material world is its flexibility, the ability to direct the efforts of its parts (subsystems), and the whole system back to where danger, to avert this danger and save themselves. To some extent this property of the systems describes Le Chatelier principle: If a system at equilibrium experiences a change then as the result of ongoing processes the equilibrium shifts to partially counter-act the imposed change .[7]

Herbaceous ecosystems (biogeocenoses) are forms of existence of living matter, specifically plant cover of the earth. They meet all the requirements of a system as multicomponent, self-regulatory "purposefully functioning structures capable of resolving problems in certain external conditions".[3] So far herbaceous ecosystems have been studied mainly on specific aspects of their functioning without sufficient generalization of accumulated data. Meanwhile, enormous factual material allows approaching its broad-scale understanding at the system level and gets "new system measurements, new genetic parameters of reality" .[4]

It is the purpose of the present communication to assess the diverse properties of biogeocenoses from a system approach perspective and to identify cause–effect relationships in the dynamics of their floristic composition under the influence of external and internal factors.

Going directly to the statement of research materials, we note that from the time of Leonardo da Vinci the main role in the formation of the soil cover along with other factors was given to higher plants including herbs.[8] Pedologists in particular L. O. Karpachevsky[9] believe that the formation of modern soil cover dates back to the Cretaceous period, when angiosperms are widely spread that is about 100 million years ago, forming deciduous forests and meadows. Perhaps modern meadows differ from those that existed 100 million years ago but their existence they are since the time and during this period meadow ecosystems developed adaptive response to external influences that threaten their existence. Thus, it is necessary to distinguish adaptation of separate types of plants and adaptation, characteristic for the entire community of these species, that is, phytocenoses as a whole. The system analysis allows you to select at least three categories of these adaptations that can be considered as factors of stability of herbaceous ecosystems. Though it may seem paradoxical, but all these factors are aimed combating against the destruction of the soil cover, the preservation of its fertility, reducing the loss of plant nutrients into the environment. And more specifically, they are directed against the negative impact on grassland cenoses primarily by large herbivores, as well as natural phenomena, causing the destruction of the sod.

The role of perennial grasses in the protection of soil from water and wind erosion is widely known.[10] However, until now, such function of herbaceous ecosystems as protection of soil from pasture erosion, that

Soil-Protecting Functions of Medicinal Plants 69

is, from destruction under the influence of large herbivorous animals, remained essentially out of sight from the scientists. In natural conditions the wild herbivorous consume on average about 10% of the biomass of natural pastures,[11] which corresponds to the general biological law of the energy pyramid.[12] However, when feeding or migration herd animals especially ungulates (buffaloes, bisons, tarpans, aurochs, antelopes, saigas, horses, and other animals) could repeatedly be situations of heavy grazing, destruction of sod, which in combinations with weather anomalies—torrential rains, hurricanes—was supposed to lead to the destruction of the soil cover, death of ecosystems. In the process of biological evolution to survive could only such communities that due to the principle of natural selection have developed adequate defense remedies primarily of soil that was the basis of existence of the plant communities and, also, grassland landscape in general. Soil as bioinert body formed under the influence of vegetation, is unable to resist the active external actions, in particular, the anthropogenesis.[13] Therefore, in the historical past, the soil could not develop appropriate protective mechanisms against the effects of major phytophages and the role of defenders of the soil from destruction in biogeocenoses was given to living beings—plants. Consider these mechanisms more with attraction of well-known facts.

The first is the reduction in the productivity of plant community with increased heavy grazing of above-ground biomass, which is designated as pasture digression. It is well known from the theory and practice of modern grassland agriculture.[14] Thus according to I. V. Larin[15] and scientists to which he refers, with increasing systematic stocking in forest-meadow area first of all disappear high and semi high perennial grasses, forming the largest biomass, common timothy, tall oat grass, awnless brome, meadow fescue, red clover, meadow foxtail. These plants towering over the other attract animals in the first place. With frequent grazing such plants rapidly waste away and fall out of the grass cover giving way to low-growing and less productive grass, such as Kentucky bluegrass, fine bent grass, white clover, etc., which without encountering competition for light and nutrition from high grass begin to dominate in the grass cover.

At the further increase of stocking the stand composition changes more rapidly due to different eatability of plants. In these cases, the plants, which are eaten most, also waste away and fall out of the grass cover. According to long-term observations, plants, grazed down 6–7 times during the summer, died or severely become sparse. On pastures remain inedible plants or plants, which are eaten not much: prostrate knotweed, silverweed, ladies'-mantle, dandelions, plantains, etc. The nutritional value of pastures for large herbivorous is falling, thus, to almost zero. The speed of pasture degradation depends on the stocking per unit of forage areas: the higher it is, the shorter the life cycle belonging to eatable species. Such dynamic change of grassland plants during high stocking from herbivores in all soil-climatic zones goes through procedure. The difference lies only in species composition of plant communities, replacing each other, and in the rate of substitution of one type by others. T. A. Rabotnov[16] pointed out that in England as a result of long unregulated grazing of sheep on pastures grew Nardus, bracken, heather, that is, they lost feed value.

The universality of the above-mentioned patterns, logically explainable by biological features of the different types of plants, from system approach should be seen as a defensive reaction of biogeocenoses from destruction of their grazing animals by decrease of stocking and in more general terms by reducing the population of herbivores in the region due to the lack of pasture forage. When the stocking of cattle on pastures decreases degraded pastures are recovered in the process of the so-called demutation.[12] However, an excessive stocking that may arise due to any reasons may not only lead to the degradation of pasture but also to pasture erosion, that is, to complete destruction of ecocenosis and destruction of the soil cover. So, in modern conditions when excessive unregulated stocking of cattle in the mountains (the Caucasus, Altai, Buryatia) sod failure (formation of paths) may exceed 60% of pasture area and soil washout from deprived of vegetation places, more than half of its power. According some data,[17] the annual reduction of soil profile in the Eastern regions of the Caucasus for this reason averaged 0.8 mm, that is, it decreased by 1 cm every 13 years. More striking manifestation of pasture erosion is observed on the Black Lands (Caspian lowland), where on winter pastures many years was converted cattle from different regions of the Northern Caucasus and Transcaucasia. As the result of excessive stocking here was almost destroyed not only the vegetation and soil, occurred desertification areas, up to the formation of drift sand.

But the decline, depression of pasture productivity is the first line of defense of grassy ecosystems. The second factor, or the mechanism of their sustainability, also directed against large herbivorous animals, triggering by the breach of the normal functioning of perennial grasses due to unfavorable weather conditions: long cold, drought, or vice versa waterlogging of the soil. In these conditions, ungulate excessive stocking on weakened plant communities may also lead to their death, sod destruction with subsequent erosion of soil cover. Protective function of herbaceous ecosystems in such situations is expressed in development in the organism of normal forage plants various substances, toxic for animals. This phenomenon is well known from literature.[18,19] First of all, this accumulation in the herbs of nitrates by a violation of the synthesis of proteins and also retardation of the growth of plants due to adverse weather conditions: hailstorm and similar anomalies. The excessive consumption of such plants causes nitrate–nitrite toxicosis of animal. It is established that feed containing more than 0.07% N-NO$_3$ is dangerous for animals and a doubling of the concentration can be fatal. In these conditions in plants of different families, including Gramineae, Fabaceae, Cyperaceae, hydrocyanic acid is formed by splitting of cyanogenic glycosides by relevant enzyme into glucose and hydrocyanic acid which is a potent poison. There are other toxicosis, which was caused by different glycosides, alkaloids, saponins, essential oils, and other substances, resulting in elevated concentrations in plants at violation of the normal processes of growth and development. For example, glycoside cumarin, which is harmless in normal conditions and found in melilots slow drying of wet habitats makes to dicumarin, highly toxic compound that causes the death of animals within 2 to 3 days. See the numerous cases of poisoning of herbivores due to eating plants affected by fungal diseases that infect weakened fodder plants. Characteristically, the most sensitive to the action of toxic compounds are the animals that came from other habitats (migrants), pregnant animal, young animals, and the animals weakened due to starvation. Often there is the death of suckling calves because of the switch to the milk toxicants contained in the forage of the cows but not rendered visible harm to the health of the latter. The toxicity of many plants depends of the habitat, the phase of development and other factors that have been studied not enough. However, it is obviously that this property was developed by fodder grasses not accidentally and, at the system level, is the protection of biocenoses from herbivores with the deterioration of habitat conditions or the need for survival in certain periods of ontogenesis. In folk and official medicine, environmental signs of the environment and phases of development of medicinal plants are also used to determine the conditions and timing of the collection of phytomass to obtain the most valuable medicinal products.

And, finally, on the last, the third obvious way to protect herbaceous ecosystems from destruction of the soil cover at the expense of weeds. It is in effect when the first two methods appear insufficient and animals, for whatever reasons, violate the integrity of the sod. It is the mass appearance of plants, usually called weeds. Among them are poisonous and indelible plants with distinct properties deter herbivores: toxicity, thorniness, hairiness, the presence of coarse stems, sharp smell, etc. To this group belong common thistle, plumeless thistle, sow thistle, black henbane, water hemlock, larkspur, aconite, white hellebore, datura— almost all the weeds of our gardens and arable lands—a total of more than 700 species, or about 15% of the floristic diversity of the natural pastures. The largest number of poisonous and noxious plants are there in the next families: Euphorbiaceae, 98% (29 species); Solanaceae, 97% (29 species); Equisetaceae, 81% (9 species); and Ranunculaceae: 54% (117 species). Quite a lot of them are in Cruciferae (Brassicaceae): 37% (60 species), Polygonaceae: 37% (39 species), and Liliaceae: 26% (34 species). In the families Gramineae (Poaceae) and Fabaceae (Papilionaceae) there are 5% (25 and 28 species, respectively), Cyperaceae: 1% (1 species).[10] Thus, from more than 1000 species of Gramineae, Fabaceae, and Cyperaceae, only 54 species are dangerous for animals, whereas in the miscellaneous herbs, which also include the weeds, there are about 700 species. Species diversity makes adequate herbage reactions to external influences depending on environmental conditions of their existence, including soil, intra- and interspecific, weather, phytosanitary, by a number of phytophages, and, as we can see, large herbivores.

In violation of the integrity of sod by ungulates animals weeds quickly fill the gaps in the grass, preventing further appearance of animals on damaged areas. Striking the adequate responses of ecosystems to the negative impact of the animals: the greater the harm caused to the grass, the sharper repellent properties of weeds are experienced by animals. In our time, clearly, this reaction can be traced about sheep enclosures, temporary stock stands of animals, where the degree of damage to turf decreases from the center of damage to the habitat periphery. One of the authors of this study had observed the emergence of a dense bed of a black henbane on the place of multiple milking in the valley of the small river in the Yaroslavl region, where the turf in the previous year was completely damaged on an area of about 100 m^2. On the periphery of the bed increased common thistles and musk thistles, and in the process of removal from the

Soil-Protecting Functions of Medicinal Plants

center of the bed their number and habitus, respectively, decreased. It is characteristic that in a year on the place of former bed was not one of the weed; ring there was only green carpet of grass, although when seeding of black henbane millions of seeds of this plant were in the soil, and this is no accident. V. R. Williams[20] described in detail the change of tall weeds consisting of various weeds, to grasses with abandonment of arable land in fallow in the steppe zone. The fallow period with prevalence of tall weeds which usually lasted one year replaced with the couch grass fallow (5–7 years), which in turn transformed into a solid fallow, consisting mainly of loose bunchgrasses (15–20 years), with a gradual transformation of the Stipa steppe with the advantage of firm bunchgrasses. In conditions of formerly widespread in the South of Russia fallow farming system the restore of natural phytocenoses, appropriate to soil and climatic conditions of the steppe, lasted, so 20–30 years. On a similar scheme, but with a different floristic composition and duration of cenogenesis procenoses (intermediate cenoses) changed on the meadows of the humid zone in case of damage or destruction of the meadow sod by animals or technical means (with anthropogenesis). For example, on a small potato field (Tver region), previously fertilized by manure, procenoses of tall weeds a year after the end of treatment consisted of wormwood—*Artemisia vulgaris* L., the following year, mainly from willow herb—*Chamerion angustifolium* L.

It weeds inhabit, first, arable land, because they are perceived by wildlife as areas with broken sod, requiring protection, and restoration of natural vegetation—grass stand or forest depending on environmental conditions. If in modern conditions people define the stocking, in the nature it regulates themselves herbaceous communities, to be exactherbaceous ecosystems, or ecosystems, including soil, and local biota. These property ecosystems have developed in the process of long evolution (cenogenesis) and natural selection of systems, ecosystems failed to produce adequate protection, have disappeared from the face of the earth as disappeared grass and soil on the Black Lands. But here in this guilty a man, against the arbitrariness of which nature was powerless. In natural condition, figuratively speaking, poison and antidote were developed simultaneously, and ecosystem natural selection happened, obviously, on the same principle as natural selection of individual species of plants or animals. Due to this in nature has been preserved dynamic equilibrium; the example of this equilibrium is the ecological balance in the system of predator–prey. The same can be said about the dynamic equilibrium that exists between herbaceous communities, on the one hand, and grazing animals, on the other. In this case as a predator are herbivores (consumers), and in the role of "prey" are fodder grasses (producers). Upon termination of grazing occurs recovery (demutation) of productive phytocenoses that used in practice by providing pasture rest. Compilation of materials on the biological productivity of natural forage lands shows that it is inversely proportional to the intensity of grazing and is a regulator to stocking.

Appearing on the places of damaged sod, weeds treat juvenile undergrowth grass almost paternal care. Otherwise you will not say. First, it is the protection of seedlings of slow-growing perennial grasses from trampling by animals, second, from their grazing in young, immature age, third, the accumulation of nutrients, especially nitrates, formed by mineralization of the destroyed sod, prevent them from losses due to leaching and denitrification, and finally, the programmed destruction of weeds order to give the living place to the next vegetation formations, that is, cereal grass, passing it "inherited" nutrients, accumulated in the plant residues.

Although in the nature after community of tall weeds would grow usually wheatgrass as a dramatic example of neutrophils, under the canopy of weeds with no less success you can grow types of forage grasses. This is evidenced by how scientific expertise and a wide practice of meadow grass cultivation. In particular, the field experiment conducted in the state farm, Voronov, of the Moscow region, on the fertile land of loam soil with coverless sowing in pure form or in mixtures of more than 10 types of cultivars of cereals and legumes, including oligotrophic, characteristic for the poor habitats (slough grass), in the year of planting white pig weed was growing abundantly. Weeding it manually on half of the area and leaving the other half before the end of vegetation did not reveal any significant difference in the condition of the herbage and yield of perennial grasses in any of subsequent years of research. These facts point to the specificity (commensalism) of relationships one-biennale weeds with perennial grasses developed under their canopy regardless of the floristic composition of each group: oligo-, meso-, or eutrophes formed on the relevant fertility soils. However, fallow of tall weeds is able, apparently, to some extent, to control the floristic composition of the procenosis that goes for a change. Research conducted in Timiryazev Moscow Agricultural Academy on the effects of the extract from the seeds of sosnovsky cow parsnip (*Herackleum sosnowskyi*) on the germination of common valerian

(*Valeriana officinalis*), St. John's wort (*Hypericum perforatum*), snow-on-the-mountain (*Euphorbia marginata*), green amaranth (*Amaranthus retroflexus*), and hare'stail grass (*Lagurus ovatus*) showed that for two species of plants extract had inhibiting effect, for two other had stimulating effect and for one, additive. It should also be that the consistent successions (changes) of procenoses (intermediate cenoses) on the ruins of the sod from tall weeds to stable (climax) phytocenosis aimed at achievement of a definite purpose—to hold and accumulate in biogeocenoses formerly accumulated elements of mineral nutrition. This can be seen in the changing attitude of plants to have in the soil mobile nutrients, in particular, nitrogen. The most demanding of them weeds such as mugwort (*Artemisia vulgaris*) white pig weed, common thistles, plumeless thistles, sow thistles, black henbane, etc. High consumption of nitrogen weeds-eutrophes indicates at least such fact as the content of crude protein, not inferior legume grasses, from 18% to 28% (calculated on the dry weight).

From the system point of view, this change of plant groupings in place of the destroyed phytocenoses can be explained any otherwise than evolutionary developed way of herbaceous ecosystems to restore the status quo with the least loss of mobile plant nutrients from the destroyed sod formed by mineralization of its organic residues. One-biennale weeds, possessing powerful starting-growth and developing the greater weight, catch nitrogen and ash elements and when death passed them to subsequent herbaceous procenoses. Already loose bunchgrasses which change rhizomatous grasses begin to inhibit the processes of mineralization of organic substances in soil that leads to a gradual accumulation of humus, and firm bunchgrasses with the prevalence of mycotrophic nutrition type complete the immobilization of nutrients transferring its main part in the organic form.

Significant role in the retention of nutrients in the soil at destruction of natural meadow turf by ungulates also plays a soil microflora. Now it is known,[21] that in untilled soil at decomposition of plant residues most immobilization mineral soil nitrogen or nitrogen fertilizers ($N-NO_3$ and $N-NH_4$) by soil microorganisms occurs in the first 10–12 days, thus preventing its infiltration losses in the underlying soil and denitrification in the form of gaseous nitrogen forms. This process is accompanied by mineralization of organic substances in soil and plant residues with the release of mineral nitrogen, which is consumed by another group of soil microflora. With the advent of vegetation on the site of the destroyed natural turf role of soil microflora in the retention of nutrients from exogenous losses are gradually decreasing. According to Smelov,[22] the number of microorganisms, mineralizing nitrogen of soil organic matter on the roots of loose bunchgrass—meadow fescue during four years of observations decreased from 9.2 to 1.4 milliards per 1 g of dry roots. Similar results were obtained with timothy. For the seventh year of life separate species of herbs accumulates from 34 to 47 tons of humus on 1 hectare transforming into immobile state 1.5–2 tons of nitrogen, from 0.5 to 1 ton of ash matter. From the results of our studies,[23] it is also obvious that consort communications in biogeocenoses between herbs microflora and soil aim to the retention of nutrients in the soil including inhibition of nitrification, and more in sabulous in comparison with loamy. With equal doses of nitrogen fertilizers and almost the same removal of nitrogen with grass yield on irrigated pasture consisted mainly of cocksfoot, and approximately equal to the content of mineral nitrogen in soil ratio nitrate form to ammonium one in the upper layer of sabulous on the average for vegetation period amounted one to five, whereas in loamy—no more than one to two. It follows from this that in sandy soil mineral nitrogen was kept from leaching by sediments in the form of a slow-moving ammonium ion ($N-NH_4$), whereas in loamy soil less susceptible to chemical suffusion, a significant part of mineral nitrogen was nitrate ($N-NO_3$). Otherwise, the environment-forming role of biocenoses was reduced to a maximum retention of mineral nitrogen in its sphere adapting to habitat nature (ecotope). Thus, even in the artificially created agrocenosis of cultural pastures under irrigation of sabulous where nitrates bigger risk of loss with infiltration waters, compared with loamy and where the best aeration, it would seem, must strengthen the processes of nitrification, the main fund of mineral nitrogen was presented ammonium form.

In general biological sense, the sustainability of ecological systems is interpreted as the ability to resist the action or to return to the initial state after exposure. This paper does not discuss factors of stability of herbaceous ecosystems associated with adaptive responses of plant communities on the change of soil, hydrological, meteorological, and other conditions, which is the subject of synecology and is described by many geobotanists.[16,22] The points of the discussion in the work are the reactions of herbaceous ecosystems produced in the process of the evolutionary development of such ecosystems and aimed at soil conservation as a primary basis of their existence.

Soil-Protecting Functions of Medicinal Plants 77

6.4 CONCLUSIONS

Opened us at the system level, this aspect of the sustainability of natural ecosystems was based on known facts, directly or indirectly pointing to the specific, essentially passive, but adequate counteraction of phytocenoses to ungulates in danger of destroying the sod and soil cover. In general relations between herbivorous animals and phytocenoses obey the laws of biological systems, "predator–prey," where the role of "predator" belongs to herbivorous, and "prey" belongs to phytocenoses. We found also the reason, functional predetermination of successive change of procenoses in the process of demutation (recovery) of destroyed vegetation which consists in evolutionary developed expediency of preservation and transfer of "inherited" moving nutrients produced when organic matter mineralization of former turf. For the first time we showed soil-protecting role of weeds in nature, contrary to the established in the agriculture opinion of them as the plunderers of soil fertility. We disclosed antientropic trend of successions herbaceous communities leading ultimately to the accumulation and structuration of mater and energy in ecosystems, ensuring their dynamic stability and, consequently, creation and preservation of soils in areas with grass vegetation.

The facts and considerations given in this paper can help researchers and practitioners to study and use medicinal plants to more reasonably plan the directions and extent of their scientific and practical activities in the field of phytotherapy and pharmacology in general.

KEYWORDS

- animals
- herbal medicine
- soil
- systemic approach
- weeds

REFERENCES

1. www.medicina.info/svedeniy/obsvedeniy.htm.
2. www.activestudy.info/zakon-arndta-schulce/.
3. Sagatovsky, V. N. *The Experience of Creation of a Categorical Apparatus of the System Approach.* Philosovskii Nauki: Moscow (*Philosoph. Sci.*, in Russian), 1976; *3*, p 75 (in Russian).
4. Afanas'ev, V. G. *System Approach and Society.* Politizdat: Moscow Publishing House of Political Literature (in Russian), 1980; p 14 (in Russian).
5. *General Biology with Principles of Historical Geology.* Vyschaya Shkola: Moscow: (Higher School, in Russian), 1980; p 4 (in Russian).
6. Chernikov, V. A.; Aleksashin A. V.; Golubev A. V. *Agroecology.* Kolos: Moscow (Ear, in Russian) Publishing House, 2000; p 14 (in Russian).
7. Glinka, N. L.; *General Chemistry.* Khimiya: Leningrad (Chemistry, in Russian) Publishing House, 1976; p 185 (in Russian).
8. Krupenikov, I. A. *The History of Pedology.* Nauka: Moscow (Science, in Russian), Publishing House, 1981; p 327 (in Russian).
9. Karpachevsky, L. O. *Soil and Pedosphere in Space Coordinates.* Agrarnaya Nauka: Moscow (Agrarian Science in Russian), 1995; *4*, pp 46–48 (in Russian).
10. Pavlovsky, E. S. *Soil-protecting Significance of Natural Forage Lands. Natural forage resources of the USSR and their use.* Nauka: Moscow (Science, in Russian), 1978; p 74–78 (in Russian).
11. Rakitnikov, A. N. *The Use of Natural Forage Resources as a Factor of Development of Agriculture.* Kolos: Moscow (Ear, in Russian) Publishing House, 1978; p 35–47 (in Russian).
12. Reymers, N. F. *Nature Management.* Mysl': Moscow (Thought, in Russian) Publishing House, 1990; p 152 (in Russian).
13. *Sustainability of soils to natural and anthropogenic influences,* Abstracts of All-Russian Conference, Dokuchaev Scientific Research Institute of Pedology: Moscow, 2002; p 489 (in Russian).
14. Andreev, N. G. *Meadland Farming.* Agropromizdat: Moscow (Publishing House of Agricultural and Industrial Literature, in Russian) Publishing House, 1985; p 83–85 (in Russian).
15. Larin, I. V. *Grassland Science and Pasture Farming.* Selhozgiz: Moscow-Leningrad (Publishing House of Agricultural Literature, in Russian) Publishing House, 1956; p 63 (in Russian).
16. Rabotnov, T. A. *Meadland Farming.* Publishing House of Moscow State University: Moscow, 1974; p 349 (in Russian).
17. Erizhev, K. A. *Mountain Hay Lands and Pastures.* Rodnik: Moscow (Spring, in Russian) Publishing House, 1998; p 320 (in Russian).
18. Vilner, A. M. *Forage Poisonings.* Kolos: Moscow (Ear, in Russian) Publishing House, 1974; p 408 (in Russian).
19. Dimitrov, S. et al. *Diagnostics of Poisoning of Animals.* Agropromizdat: Moscow (Publishing House of Agricultural and Industrial Literature, in Russian) Publishing House, 1986; p 284 (in Russian).

20. Williams, V. R. *Collected Papers*. Selhozgiz: Moscow (Publishing House of Agricultural Literature, in Russian) Publishing House, 1949; *3*, p 132–135 (in Russian).
21. Owen, O. S. The Protection of Natural Resources. Kolos: Moscow (Ear, in Russian) Publishing House, 1977; p 179–180 (in Russian).
22. Smelov, S. P. *Theoretical Foundations of Grassland Science*. Kolos: Moscow (Ear, in Russian) Publishing House, 1966; p 121–126 (in Russian).
23. Afanas'ev, R. A. Fertilizer of Intensive Irrigated Pastures in the Nonchernozemic Zone of the RSFSR. Summary of the Ph.D. Thesis, Latvian Institute of Agriculture and Rural Economy: Scriveri, 1987; p 44 (in Russian).

CHAPTER 7

Glycosylation of Polyphenols in Tannin-Rich Extracts from *Euphorbia antisyphilitica, Jatropha dioica*, and *Larrea tridentata*

JANETH VENTURA-SOBREVILLA[1], GERARDO GUTIÉRREZ-SÁNCHEZ[2], CARL BERGMANN[2], PARASTOO AZADI[2], DANIEL BOONE-VILLA[1], RAUL RODRIGUEZ[1], and CRISTÓBAL N. AGUILAR[1*]

[1]*Food Research Department, School of Chemistry, Universidad Autónoma de Coahuila, PO Box 252, 25000 Saltillo, Coahuila, México*

[2]*Complex Carbohydrate Research Center, The University of Georgia, Athens, GA 33602-4712, USA*

Corresponding author. E-mail: cristobal.aguilar@uadec.edu.mx

ABSTRACT

Potency of activity and efficiency of polyphenols-rich plant extracts to exhibit several biological properties are highly influenced by the chemical interactions of the biomolecules and physicochemical conditions of itself system, and possibly it is more severe in tannin-rich plant extracts due their complexity and reactivity, for this reason is necessary to define the class of components with high influence in the bioactivity of the phytochemicals. The objective of this study was to analyze the composition of carbohydrates present in crude and fractionated extracts of these plants using gas chromatography–coupled mass spectrometry (GC–MS). A wide gamma of carbohydrates was found in extracts from *Euphorbia antisyphilitica* Zucc., *Jatropha dioica* Cerv., and *Larrea tridentata* Cov. Arabinose, rhamnose, fucose, xylems, glucoronic acid, galacturonic acid, mannose, galactose, and glucose were detected by GC–MS, but only arabinose, rhamnose, and glucose were consistently observed in all crude and fractionated samples, revealing the necessity for understanding their role, function, and structure sugar phenolic compounds in foods and medicinal plants.

7.1 INTRODUCTION

Tannins are present in fruits and vegetables and they are considered secondary plant metabolites. Tannins are the second most abundant group of plant phenolic compounds just after lignins (Bhat et al., 1998). The medicinal properties of tannins, such as antimicrobial, antiviral, antitumor, anticarcinogen, antioxidant, and anti-inflammatory have been recently reviewed and highlighted (Ascacio-Valdés et al., 2011; Viuda-Martos et al., 2010; Isemura et al., 1999). On the basis of their structural characteristics, it is possible to divide them into four major groups: gallotannins, ellagitannins, complex tannins, and condensed tannins (Ascacio-Valdes et al., 2011). Gallotannins and ellagitannins have a polyol core, commonly sugar, esterified by galloyl units, whereas in complex tannins catechin unit is bound glycosidically to gallotannin or ellagitannin units (Ascacio-valdés et al., 2011; Khanbabaee and Ree, 2001).

Phenolic compounds are highly unstable and easily to degradation through factors, such as light, temperature, pH, among others. In some studies has been reviewed that the polyphenol compounds are linked by o-glycosidic bonds on sugar molecule in order to increase their stability (Castañeda-Ovando et al., 2011). Actually, there are no reports about sugar present in the rich phenolic extract from candelilla (*Euphorbia antisyphylitica* Zucc.), creosote bush (*Larrea tridentata* Cov.), and dragon's blood (*Jatropha dioica* Cerv.), which are plant species widely distributed in the semiarid regions of southwest of the United States and Northern Mexico. The leaves, roots, and stems of these plants have been used to prepare infusions or tea in the traditional medicine to treat oral, skin, digestive, liver, respiratory, and venereal diseases, acne and diabetes, also have been reported some antimicrobial, antiviral, antiseptic, and antioxidant properties (Martins et al., 2013; Fresnedo-Ramírez and Orozco-Ramírez, 2012; Mendez et al., 2012; Osorio et al., 2010; Saucedo-pompa et al., 2009; Fresnedo-Ramírez and Orozco-Ramírez, 2012; Vargas-Arispuro et al., 2005; Martínez, 1992).

Earlier, our group reported that extract of these plants are good source of monomer and polymeric phenolic compounds, particularly tannins (Rojas-Molina, 2013; Ascacio-Valdéz et al., 2013; Ruiz-Martínes et al., 2011). In addition, these extracts contain free and linked to polyphenol sugars so, have been used as carbon source or support for design of culture medium to produce lees complexes antioxidant molecules and several enzymes by solid and liquid fermentation (Belmares et al., 2009; Ventura et al., 2009). The aim of this study was to characterize the sugar monomers presents in phenolic extracts of the three native plants of Mexican desert mentioned above.

7.2 MATERIALS AND METHODS

7.2.1 PLANT MATERIAL

Candelilla, creosote bush, and dragon's blood were collected from the Southern suburban zone of Saltillo City, Coahuila, Mexico in April and May, 2012. The plant materials were transported to the Food Research Department in School of Chemistry at the Universidad Autónoma de Coahuila, after which the plants were washed with distilled water and dried at 60°C and pulverized in a manual mill.

7.2.2 PHYSICOCHEMICAL CHARACTERIZATION

Moisture, total solid, ash, fat, protein, crude fiber were determined by following the official procedures of AOAC (Horwitz, 2000). Total sugars were determined using the colorimetric method reported by DuBois et al., (1956) and reducing sugar content was evaluated using the method of dinitrosalysilic acid (Miller, 1989). Total hydrolysable and condensed tannins were determined using the method reported by Ventura et al. (2008).

7.2.3 EXTRACT PREPARATION

Powdered plant samples (100 g) were mixed with 400 mL of 70% aqueous acetone in flasks covered with aluminium foil to avoid light exposure. The extraction process was carried out by reflux distillation for 12 h at 40°C.

After this process, the samples were filtered using Whatman filter paper no. 4 and centrifuged at 1360 g for 15 min. The acetone was removed with a rotary evaporator (YAMATO, RE540) at less than 60°C and avoiding light exposure. The crude extracts were freeze-dried and transported to the Complex Carbohydrate Research Center of the University of Georgia, Athens, GA, USA.

7.2.4 CRUDE EXTRACT FRACTIONATION

The freeze-dried samples (0.3 g) were partially homogenized with deionized water (15 mL) and were then filtered through a Whatman glass microfilter (FG/A). The aqueous extracts were fractionated using an amberlite XAD-16 column. The amberlite was prepared for chromatography by prewashing in ethanol then pre-equilibrating in water for 12–14 h. The crude extract was absorbed onto a vacuum-aspirated column of amberlite XAD-16 resin. Each column was eluted with deionized water (50 mL) until the sugary pale elutate was uncolored. The less polar compound absorbed were eluted with ethanol until a colorless liquid was acquired. The ethanol was removed with a stream of helium using an evaporating unit (REACTI-VAP). To reduce the evaporation time, the samples were placed into a water bath at 40°C. Then, the samples were dissolved in deionized water and the aqueous and phenolic fractions were freeze-dried for further analysis.

7.2.6 SUGAR STANDARD PREPARATION

All reagents were ultra pure grade. Standard No. 1 contained (1 mg/mL) of arabinose, fucose, galacturonic acid, glucose, N-acetyl-glucosamine, and N-acetyl-mannosamine. Standard No. 2 contained (1 mg/mL) rhamnose, xylose, glucoronic acid, galactose, mannose, and N-acetyl-galactosamine.

7.2.7 HYDROLYSIS, METHYLATION, ACETYLATION, AND TRISILATION

Before hydrolysis, 0.4–0.8 mg of powdered sample was placed inside screw cap tubes with 20 μL of myoinositol (1 mg/mL). The samples and

Glycosylation of Polyphenols in Tannin-Rich Extracts 85

standards were frozen and lyophilized until dry. There were hydrolyzed and methylated with 1 mL of anhydrous methanolic HCl (36.5 mg/mL) and incubated at 80°C for 16–18 h. Solvent was evaporated under a stream of nitrogen. Two evaporations with methanol were followed by dissolving the sample in 200 µL of methanol and evaporating the methanol with nitrogen. For acetylation, 200 mL of methanol, 100 µL of pyridine and 100 µL of acetic anhydride were added to the sample and the tubes were placed at room temperature for 30 min. Methanol was removed and 400 µL of Tris-Sil Z were added. The samples were incubated at 8°C for 20 min and partially dried (less than 150 µL) with a stream of nitrogen at room temperature and mixed with 2 mL of hexane. The samples then were filtered through a glass wool packed Pasteur pipette and dried down. Finally, the samples were dissolved in 100 µL of hexane and analyzed by gas chromatography–mass spectrometry (GC–MS).

7.2.8 GAS CROMATOGRAPHY–MASS SPECTROMETRY ANALYSIS

Analysis was performed using a GC (Agilent Technologies 6890N) coupled mass spectrometer (Agilent Technologies 5975B). Sample separations were carried out on a capillary column of silica gel Supelco EC-1 (30 m × 0.25 mm) and following the temperature programme: 2 min at an initial temperature of 80°C, increased to 170°C at 3°C per min, then to 250°C at 4°C per min and kept for 5 min at 250°C. Injector temperature was kept at 250°C and 280°C for the refraction index detector. A volume of 10 µL was used. The identification of mono sugars from crude and fraction samples was based on retention times and main fragments of standard monosaccharide peaks.

7.3 RESULTS

7.3.1 PROXIMAL CHEMICAL COMPOSITION

Table 7.1 shows chemical characterization of plant materials. Sugar expressed as soluble and reducing sugar and crude fiber were the most abundant constituents of these plants, therefore, it is possible and feasible to analyze the kind of monosacharides are presents. On the other hand, the content in total polyphenols was higher in *L. tridentata* and *E.*

86 Green Chemistry and Biodiversity: Principles, Techniques, and Correlations

antisyphilitica than *J. dioica*. In *L. tridentata* several phenolic compounds such a as ellagic acid, gallic acid, and catechin (Rojas-Molina et al., 2013; Aguilera-Carbo et al., 2008; Ventura et al., 2009), candelitannin (Ascacio-Valdés et al., 2013) have been reported for these plants.

TABLE 7.1 Proximal Chemical Composition of *E. antisyphilitica, L. tridentata,* and *J. dioica* Expressed as g Per 100 g of Dry Plant (%).

Components	*E. antisyphilitica*	*L. tridentata*	*J. dioica*
Moisture	4.44	5.9	6.22
Total solids	95.60	94.1	93.70
Ash	10.89	8.0	11.23
Soluble sugars	23.93	9.3	29.11
Reducing sugars	15.62	9.1	25.60
Protein content	1.42	4.1	2.35
Fat content	15.92	4.3	2.50
Crude fiber	9.05	18.8	25.74
Total polyphenols	7.20	9.6	1.25
Condensed tannins	1.60	8.0	1.95
Hydrolysable tannins	5.50	1.6	2.80

7.3.2 SUGAR ANALYSIS FROM E. antisyphilitica

E. antisyphilitica crude (EC), *E. antisyphilitica* aqueous fraction (EAF), and *E. antisyphilitica* ethanolic fraction (EEF) were analyzed by gas chromatography and mass spectrometry, and the sugar compounds were confirmed by comparing their retention times and MS data to those of authenticated standards (Table 7.2). Thus, presence of arabinose, rhamnose, xylose, mannose, galactose, glucose, and glucuronic acid were in EC, EAF, and EEF, but fucose and galacturonic acid were only found in EAF. Glucose was the most abundant sugar present EC (89.5%), EAF (77.3%), and EEF (92.8%). In addition, several no-sugar compounds with mass spectra similar to inositol were detected.

Glycosylation of Polyphenols in Tannin-Rich Extracts 87

TABLE 7.2 Retention Times and Main Fragments in GC–MS of Sugar Monomers from *E. antisyphilitica.*

Samples	Retention time (min)	Main fragments (*m/z*)	Compounds	Content (%)
E. antisyphylitica crude (EC)	12.394	73, 133, 147, 204, 217	Ara	3.3
	12.745	73, 133, 147, 204, 217	Ara	
	13.404	73, 133, 147, 204, 217	Rha	4.1
	13.719	73, 133, 147, 204, 217	Rha	
	15.658	73, 133, 147, 204, 217	Xyl	1.6
	16.421	73, 133, 147, 204, 217	Xyl	
	21.181	73, 133, 147, 204, 217	Man	0.2
	22.768	73, 133, 147, 204, 217	Gal	0.5
	25.028	73, 133, 147, 204, 217	Glc	89.5
	26.072	73, 133, 147, 204, 217	Glc	
	25.610	73, 133, 147, 204, 217	GlcA	0.9
E. antisyphylitica aqueous fraction (EAF)	12.403	73, 133, 147, 204, 217	Ara	9.1
	12.715	73, 133, 147, 204, 217	Ara	
	13.408	73, 133, 147, 204, 217	Rha	3.5
	13.719	73, 133, 147, 204, 217	Rha	
	14.181	73, 133, 147, 204, 217	Fuc	0.5
	14.924	73, 133, 147, 204, 217	Fuc	
	15.687	73, 133, 147, 204, 217	Xyl	4.4
	16.431	73, 133, 147, 204, 217	Xyl	
	18.5	73, 133, 147, 204, 217, 230	GlcA	1.2
	25.630	73, 133, 147, 204, 217, 234	GlcA	
	21.181	73, 133, 147, 204, 217	Man	0.4
	21.512	73, 133, 147, 204, 217	Gal	3.1
	22.768	73, 133, 147, 204, 217	Gal	
	23.431	73, 147, 217, 234	GalA	0.5
	25.017	73, 133, 147, 204, 217	Glc	77.3
	26.062	73, 133, 147, 204, 217	Glc	

TABLE 7.2 *(Continued)*

Samples	Retention time (min)	Main fragments (*m/z*)	Compounds	Content (%)
E. antisyphylitica ethanolic fraction (EEF)	12.404	73, 133, 147, 204, 217	Ara	1.5
	12.725	73, 133, 147, 204, 217	Ara	
	13.408	73, 133, 147, 204, 217	Rha	2.7
	13.719	73, 133, 147, 204, 217	Rha	
	15.688	73, 133, 147, 204, 217	Xyl	1.5
	16.431	73, 133, 147, 204, 217	Xyl	
	21.181	73, 133, 147, 204, 217	Man	0.7
	22.768	73, 133, 147, 204, 217	Gal	0.5
	25.018	73, 133, 147, 204, 217	Glc	92.8
	26.062	73, 133, 147, 204, 217	Glc	
	25.630	73, 133, 147, 204, 217, 234	GlcA	0.3

Ara, arabinose; Fuc, fucose; Gal, galactose; GalA, galacturonic acid; Glc, glucose; GlcA, glucuronic acid; Man, mannose; Rha, rhamnose; Xyl, xylose.

% Content expresses the amount of monosaccharide respect to all sugar detected.

7.3.3 SUGAR ANALYSIS FROM L. tridentata

The GC–MS study of the *L. tridenata* crude (LC), *L. tridenata* aqueous fraction (LAF), and *L. tridenata* ethanolic fraction (LEF) fraction showed that the main sugar compound is glucose 74.1%, 79.8%, and 81.6%, respectively (Table 7.3). The fractionating of freeze-dried tannin extract from creosote bush led to detect mannose in LAF and LEF. Galactose was identified in LAF but not in LEF extract, whereas arabinose, rhamnose, and mannose were found in the crude and fractionated sample from creosote bush.

All *Larrea* samples revealed the presence of a single large peak with a retention time of 23.8 min and mass spectrum similar to inositol, it suggests that polyols different to sugar are related to phenolic compounds.

Glycosylation of Polyphenols in Tannin-Rich Extracts 89

TABLE 7.3 Retention Times and Main Fragments in GC–MS of Sugar Monomers from *L. tridentata*.

Samples	Retention time (min)	Main fragments (*m/z*)	Compounds	Content (%)
L. tridentata crude (LC)	12.414	73, 133, 147, 204, 217	Ara	13.0
	12.725	73, 133, 147, 204, 217	Ara	
	13.418	73, 133, 147, 204, 217	Rha	9.3
	22.768	73, 133, 147, 204, 217	Gal	3.6
	24.094	73, 133, 147, 204, 217	Gal	
	25.018	73, 133, 147, 204, 217	Glc	74.1
	26.072	73, 133, 147, 204, 217	Glc	
L. tridentata aqueous fraction (LAF)	12.404	73, 133, 147, 204, 217	Ara	9.5
	12.711	73, 133, 147, 204, 217	Ara	
	13.408	73, 133, 147, 204, 217	Rha	5.2
	13.709	73, 133, 147, 204, 217	Rha	
	21.181	73, 133, 147, 204, 217	Man	1.6
	22.758	73, 133, 147, 204, 217	Gal	3.9
	25.007	73, 133, 147, 204, 217	Glc	79.8
	26.052	73, 133, 147, 204, 217	Glc	
L. tridentata ethanolic fraction (LEF)	12.404	73, 133, 147, 204, 217	Ara	11.1
	12.705	73, 133, 147, 204, 217	Ara	
	13.408	73, 133, 147, 204, 217	Rha	6.9
	13.719	73, 133, 147, 204, 217	Rha	
	21.163	73, 133, 147, 204, 217	Man	0.4
	25.008	73, 133, 147, 204, 217	Glc	81.6
	26.062	73, 133, 147, 204, 217	Glc	

Ara, arabinose; Fuc, fucose; Gal, galactose; GalA, galacturonic acid; Glc, glucose; GlcA, glucuronic acid; Man, mannose; Rha, rhamnose; Xyl, xylose.

% Content expresses the amount of monosaccharide respect to all sugar detected.

7.3.3 SUGAR ANALYSIS FROM J. dioica

Retention times and main fragments from *J. dioica* crude (JC), *J. dioica* aqueous fraction (JAF) and *J. dioica* ethanolic fraction (JEF) are concentrated in Table 7.4. The results demonstrated that all the samples contain

90 Green Chemistry and Biodiversity: Principles, Techniques, and Correlations

arabinose, rhamnose, xylose, glucuronic acid, mannose, galactose, and glucose with different ratios (Figure 7.1). The glucose was the most abundant sugar in all *J. dioica* samples; however, JC sample showed lower percentage than JAF and JEF. Glucuronic acid was no detected in JC.

TABLE 7.4 Retention Times and Main Fragments in GC–MS of Sugar Monomers from *J. dioica.*

Samples	Retention time (min)	Main fragments (m/z)	Compounds	Content (%)
J. dioica crude (JC)	12.424	73, 133, 147, 204, 217	Ara	11.6
	12.719	73, 133, 147, 204, 217	Ara	
	13.418	73, 133, 147, 204, 217	Rha	6.9
	13.709	73, 133, 147, 204, 217	Rha	
	15.684	73, 133, 147, 204, 217	Xyl	10.8
	16.431	73, 133, 147, 204, 217	Xyl	
	19.404	73, 133, 147, 294, 217	GalA	5.5
	23.421	73, 133, 147, 204, 217, 234	GalA	
	23.725	73, 147, 217, 234	GalA	
	21.191	73, 133, 147, 204, 217	Man	1.2
	22.778	73, 133, 147, 204, 217	Gal	6.6
	24.092	73, 133, 147, 204, 217	Gal	
	25.018	73, 133, 147, 204, 217	Glc	57.5
	26.062	73, 133, 147, 204, 217	Glc	
J. dioica aqueous fraction (JAF)	12.373	73, 133, 147, 204, 217	Ara	0.3
	12.695	73, 133, 147, 204, 217	Ara	
	13.408	73, 133, 147, 204, 217	Rha	0.5
	15.682	73, 133, 147, 204, 217	Xyl	0.8
	16.431	73, 133, 147, 204, 217	Xyl	
	18.513	73, 133, 147, 204, 217, 230	GlcA	1.2
	25.628	73, 133, 147, 204, 217, 234	GlcA	
	19.393	73, 133, 147, 204, 217, 234	GalA	5.3
	23.401	73, 133, 147, 204, 217, 234	GalA	
	23.744	73, 133,147, 204, 217, 234	GalA	
	21.191	73, 133, 147, 204, 217	Man	3.2

Glycosylation of Polyphenols in Tannin-Rich Extracts

TABLE 7.4 *(Continued)*

Samples	Retention time (min)	Main fragments (*m/z*)	Compounds	Content (%)
	22.768	73, 133, 147, 204, 217	Gal	2.8
	24.092	73, 133, 147, 204, 217	Gal	
	25.008	73, 133, 147, 204, 217	Glc	85.6
	26.062	73, 133, 147, 204, 217	Glc	
J. dioica ethanolic fraction (JEF)	12.393	73, 133, 147, 204, 217	Ara	7.0
	12.705	73, 133, 147, 204, 217	Ara	
	13.428	73, 133, 147, 204, 217	Rha	7.7
	13.709	73, 133, 147, 204, 217	Rha	
	14.181	73, 133, 147, 204, 217	Fuc	0.1
	15.687	73, 133, 147, 204, 217	Xyl	3.6
	16.431	73, 133, 147, 204, 217	Xyl	
	18.520	73, 133, 147, 204, 217, 230	GlcA	0.9
	25.640	73, 133, 147, 204, 217, 234	GlcA	
	19.403	73, 133, 147, 204, 217	GalA	0.7
	23.411	73, 133, 147, 204, 217, 230	GalA	
	23.772	73, 133,147, 204, 217, 230	GalA	
	21.181	73, 133, 147, 204, 217	Man	0.6
	22.768	73, 133, 147, 204, 217	Gal	0.5
	25.058	73, 133, 147, 204, 217	Glc	78.9
	26.082	73, 133, 147, 204, 217	Glc	

Ara, arabinose; Fuc, fucose; Gal, galactose; GalA, galacturonic acid; Glc, glucose; GlcA, glucuronic acid; Man, mannose; Rha, rhamnose; Xyl, xylose.
% Content expresses the amount of monosaccharide with respect to all sugar detected.

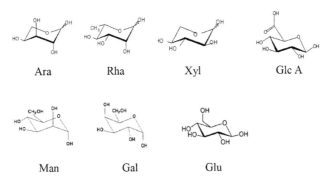

FIGURE 7.1 Glycosides present in tannin-rich extracts of Mexican desert plants.

7.4 DISCUSSION

We analyzed the profile of pentoses and hexoses present in acetone crude extract and fractionated aqueous and ethanolic extract of three plants from northeast of Mexico used in traditional medicinal.

The fractionation of crude extract was performed on amberlite XAD-16, this resin is used to discard undesirable compounds when is eluted with water, and the ethanol elution are useful to obtain rich polyphenols extract fractionated (Ascacio-Valdés et al., 2010; Ascacio-Valdés et al., 2013; Seeram et al., 2005).

Several studies had investigated the influence of different sugar addition on anthocyanin content and stability, and theirs results showed that sugar namely, glucose, saccharose, and trehalose had an positive effect on theses phenolic compounds (Kopjar et al., 2012; Lewis et al., 1995; Wrolstad et al., 1990). Ludwig et al. (2013) found that the sugar addition during roasting process of coffee increases the antioxidant capacity. Kopjar et al. (2009) studied the influence of addition of sugar mixtures on phenolics compounds content of blackberries during freeze storage for 1 year, they found that samples with sugar had higher phenolic contents than samples without sugars.

Copigmentation is a phenomenon in which phenolics compounds form molecular association with several substances called cofactors to generate more stable complexes. Intermolecular copigmentation takes place when sugars are used as cofactors (Castañeda-Ovando et al., 2009; Cavalcanti et al., 2011).

7.5 CONCLUSION

The present paper gives information about the principal sugars present in phenolics extracts from *E. antisyphilitica* Zucc, *J. dioica* Cerv., and *L. tridentata* Cov. Glucose was most abundant sugar presents in all the samples. The fractionation of the crude extract allowed identifying some sugar such as fucose, glucuronic acid, and galacturonic acid in several sugar and phenolic fractions. Finally, GC–MS was useful for analyzing the sugar content and the study showed that are many the sugar related with the polyphenols but also there are no sugar polyols that need to be investigated.

ACKNOWLEDGMENTS

The authors are thankful to the Mexican Council of Science and Technology (CONACYT) for financial support through the project, CONAFOR-CONACYT C-01-91633, and the postgraduate fellowship provided to J. M. Ventura-Sobrevilla.

KEYWORDS

- *Euphorbia antisyphilitica*
- *Jatropha dioica*
- *Larrea tridentata*
- phenolic extract
- sugar composition

REFERENCES

Aguilera-Carbo, A. F.; Augur, C.; Prado-Barragan, L. A.; Aguilar, C. N.; Favela-Torres, E. Extraction and Analysis of Ellagic Acid from Novel Complex Sources. *Chem. Pap.* **2008,** *62* (4), 440–444. DOI: 10.2478/s11696-008-0042-y.

Ascacio-Valdés, J. A.; Aguilera-Carbó, A.; Martínez-Hernández, J. L.; Rodríguez-Herrera, R.; Aguilar, C. N. *Euphorbia antisyphilitica* Residues as a New Source of Ellagic Acid. *Chem. Pap.* **2010,** *64* (4), 528–532. DOI:10.2478/s11696-010-0034-6.

Ascacio-valdés, J. A.; Buenrostro-figueroa, J. J.; Aguilera-carbo, A.; Prado-Barragan, A.; Rodríguez-Herrera, R.; Aguilar, C. N. Ellagitannins : Biosynthesis , Biodegradation and Biological Properties. *J. Med. Plants Res.* **2011,** *5* (19), 4696–4703.

Ascacio-Valdés, J.; Burboa, E.; Aguilera-Carbo, A. F.; Aparicio, M.; Pérez-Schmidt, R.; Rodríguez, R.; Aguilar, C. N. Antifungal Ellagitannin Isolated from *Euphorbia antisyphilitica* Zucc. *Asian Pac. J. Trop. Med.* **2013,** *3* (1), 41–46.

Belmares, R.; Garza, Y.; Rodríguez, R.; Contreras-Esquivel, J. C.; Aguilar, C. N. Composition and Fungal Degradation of Tannins Present in Semiarid Plants. *Elect. J. Environ. Agri. Food Chem.* **2009,** *8* (4), 312–318. Retrieved from http://www.ncbi.nlm.nih.gov/pubmed/19943419.

Bhat, T. K.; Singh, B.; Sharma, O. P. Microbial Degradation of Tannins—a current Perspective. *Biodegradation* **1998,** *9* (5), 343–57. Retrieved from http://www.ncbi.nlm.nih.gov/pubmed/10192896.

Castañeda-Ovando, A.; Pacheco-Hernández, M. D. L.; Páez-Hernández, M. E.; Rodríguez, J. A.; Galán-Vidal, C. A. Chemical Studies of Anthocyanins: A Review. *Food Chem.* **2009,** *113* (4), 859–871. DOI: 10.1016/j.foodchem.2008.09.001.

Cavalcanti, R. N.; Santos, D. T.; Meireles, M. A. a.Non-thermal stabilization Mechanisms of Anthocyanins in Model and Food Systems—An Overview. *Food Res. Int.* **2011,** *44* (2), 499–509. DOI: 10.1016/j.foodres.2010.12.007.

DuBois, M.; Gilles, K. A.; Hamilton, J. K.; Rebers, P. A.; Smith, F. Colorimetric Method for Determination of Sugars and Related Substances. *Anal. Chem.* **1956,** *28* (3), 350–356. DOI: 10.1021/ac60111a017.

Fresnedo-Ramírez, J.; Orozco-Ramírez, Q. Diversity and Distribution of Genus Jatropha in Mexico. *Genet. Resour. Crop Evol.* *60* (3), 1087–1104 **2012,** DOI: 10.1007/s10722-012-9906-7.

Horwitz, W.; Ed. *Official Methods of Analysis of AOAC International*, 17th ed; Association of Official Analytocal Chemists: Gaithersburg, Maryland, 2000.

Isemura, M.; Saeki, K.; Minami, T.; Hayakawa, S.; Kimura, T.; Shoji, Y.; Sazuka, M. Inhibition of Matrix Metalloproteinases by Tea Catechins and Related Polyphenols. *Ann. New York Acad. Sci.* **1999,** *878*, 629–31. Retrieved from http://www.ncbi.nlm.nih.gov/pubmed/10415792.

Khanbabaee, K.; Ree, T. Van. Tannins : Classification and Definition. *Nat. Prod. Rep.* **2001,** *18*, 641–649. DOI: 10.1039/b101061l.

Kopjar, M.; Jakšić, K.; Piližota, V. Influence of Sugars and Chlorogenic Acid Addition on Anthocyanin Content, Antioxidant Activity, and Color of Blackberry Juice During Storage. *J Food Process. Preser.* **2012,** *36* (6), 545–552. DOI: 10.1111/j.1745-4549.2011.00631.x.

Kopjar, M.; Tiban, N. N.; Pilizota, V.; Babic, J. Stability of Anthocyanins, Phenols, and Free Radical Scavenging Activity Through Sugar Addition During Frozen Storage of Blackberries. *J. Food Process. Preser.* **2009,** *33* (2009), 1–11. DOI: 10.1111/j.1745-4549.2008.00244.x.

Ludwig, I. A.; Bravo, J.; Paz De Peña, M.; Cid, C. Effect of Sugar Addition (torrefacto) During Roasting Process on Antioxidant Capacity and Phenolics of Coffee. *LWT - Food Sci. Technol.* **2013,** *51* (2), 553–559. DOI: 10.1016/j.lwt.2012.12.010.

Martínez, M. *Las Plantas Medicinales de México*, 6th ed; Botas: México, p 656, 1992.

Martins, S.; Amorim, E. L. C.; Peixoto, T. J. S.; Saraiva, A. M.; Pisciottano, M. N. C.; Aguilar, C. N.; Teixeira, J. A., et al. Antibacterial Activity of Crude Methanolic Extract and Fractions Obtained from Larrea tridentata Leaves. *Ind. Crops Prod.* **2013,** *41*, 306–311.

Miller, L. G. Use of Dinitrosalicylic Acid Reagent for Determination of Reducing Sugar. *Anal. Chem.* **1989,** *31*, 426–428.

Osorio, E.; Flores, M.; Hernández, D.; Ventura, J.; Rodríguez, R.; Aguilar, C. N. Biological Efficiency of Polyphenolic Extracts from Pecan Nuts Shell (*Carya Illinoensis*), Pomegranate Husk (*Punica granatum*), and Creosote Bush Leaves (*Larrea tridentata* Cov.) Against Plant Pathogenic Fungi. *Ind. Crops Prod.* **2010,** *31,* 153–157. DOI: 10.1016/j.indcrop.2009.09.017.

Ruiz-Martínes, J.; Ascacio, J. A.; Rodríguez, R.; Morales, D.; Aguilar, C. N. Phytochemical Screening of Extracts from some Mexican Plants used in Traditional Medicine. *J. Med. Plants Res.* **2011,** *5* (13), 2791–2797.

Saucedo-pompa, S.; Rojas-molina, R.; Aguilera-carbó, A. F.; Saenz-galindo, A.; La, H. De, Jasso-cantú, D.; Aguilar, C. N. Edible Film Based on Candelilla Wax to Improve the Shelf Life and Quality of Avocado. *Food Res. Int.* **2009,** *42,* 511–515. DOI: 10.1016/j.foodres.2009.02.017.

Glycosylation of Polyphenols in Tannin-Rich Extracts 95

Seeram, N.; Lee, R.; Herber, D. Rapid Large Scale Purification of Ellagitannins from Pomegranate Husk, a Byproduct of the Commercial Juice Industry. *Sep.Purif. Technol.* **2005**, *41*, 49–55. DOI: 10.1016/j.seppur.2004.04.003.

Vargas-Arispuro, I.; Reyes-Báez, R.; Rivera-Castañeda, G.; Martínez-Téllez, M. A.; Rivero-Espejel, I. Antifungal Lignans from the Creosotebush (*Larrea tridentata*). *Ind. Crops Prod.* **2005**, *22* (2), 101–107. DOI: 10.1016/j.indcrop.2004.06.003.

Ventura, J, Gutiérrez-Sanchez, G.; Rodríguez-Herrera, R.; Aguilar, C. N. Fungal Cultures of Tar Bush and Creosote Bush for Production of Two Phenolic Antioxidants (pyrocatechol and gallic acid). *Folia Microbiol.* **2009**, *54* (3), 199–203. DOI: 10.1007/s12223-009-0031-8.

Ventura, Janeth, Belmares, R.; Aguilera-carbo, A.; Gutiérrez-Sanchez, G.; Rodríguez-Herrera, R.; Aguilar, C. N. Fungal Biodegradation of Tannins from Creosote Bush (*Larrea tridentata*) and Tar Bush (*Fluorensia cernua*) for Gallic and Ellagic Acid Production. *Food Technol. Biotechnol.* **2008**, *46* (2), 213–217.

Viuda-Martos, M.; Fernández-López, J.; Pérez-Álvarez, J. A. Pomegranate and its Many Functional Components as Related to Human Health: A Review. *Comp. Rev. Food Sci. Food Saf.* **2010**, *9* (6), 635–654. DOI: 10.1111/j.1541-4337.2010.00131.x.

Wrolstad, R. E.; Skrede, G.; Lea, P.; Enersen, G. Influence of Sugar on Anthocyanin Pigment Stability in Frozen Strawberries. *J. Food Sci.* **1990**, *55* (4), 1064–1065. DOI: 10.1111/j.1365-2621.1990.tb01598.x.

CHAPTER 8

Analysis and Quantification of *Larrea tridentata* Polyphenols Obtained by Reflux and Ultrasound-Assisted Extraction

KARINA CRUZ-ALDACO[1], DANIELA SÁNCHEZ-ALDANA[2], SALVADOR ORTEGA-HERNÁNDEZ[2], GUADALUPE CÁRDENAS-FÉLIX[2], ANTONIO AGUILERA-CARBO[3], JUAN ALBERTO ASCASIO-VALDES[1], RAÚL RODRIGUEZ-HERRERA[1], and CRISTÓBAL NOÉ AGUILAR[1*]

[1]*Food Research Department, School of Chemistry, Universidad Autónoma de Coahuila, Blvd. Venustiano Carranza S/N Col. República Oriente, 25280 Saltillo, Coahuila, México*

[2]*Innovation and Technological Development Center, BAFAR Group, Km 7.5 Carretera a Cuauhtémoc S/N Col. Las ánimas, 31450 Chihuahua, Chihuahua, Mexico*

[3]*Department of Animal Nutrition, Universidad Autónoma Agraria Antonio Narro, Calzada Antonio Narro 1923, Col. Buenavista, 25315 Saltillo, Coahuila, México*

[*]*Corresponding author. E-mail: cristobal.aguilar@uadec.edu.mx*

ABSTRACT

Ultrasound-assisted extraction was evaluated as a simpler and more effective alternative to conventional extraction methods for the isolation of polyphenols from leaves of gobernadora (*Larrea tridentata* Cov.). The gobernadora samples were extracted with water and ethanol, under conditions of sonication. The ultrasonic extraction was compared with the conventional method of refluxing boiling solvents in a Soxhlet extractor, on the yields of both the total polyphenols recovered and identified by high-performance liquid chromatography coupled with mass spectrometry. Other parameters quantified were hydrolyzable tannins, condensed tannins, and percentage of antioxidant activity (1,1-diphenyl-2-picrylhydrazyl). It was found an increment of both condensed and hydrolyzable tannins with an increase of ethanol concentration with a low ultrasound-assisted extraction is an excellent alternative for recovery of polyphenols of gobernadora. The increase in the ethanol concentration is proportional to the increase of the hydrolyzable polyphenols and condensed tannins. Analysis of High performance liquid chromatography/mass spectrometry (HPLC/MS) allowed to identify 10 mass corresponding to compounds as coumaric acid, catechins, and lignans were identified. The most important compounds found was nordihydroguaiaretic acid.

8.1 INTRODUCTION

Larrea tridentata (Sesse & Mocino ex DC.) Coville is also known as Larrea, chaparral, or creosote bush in the United States and gobernadora and hediondilla (little smelly one) in Mexico. It is a shrubby plant belonging to the family *Zygophyllaceae,* which dominates some areas of the southwest desert in the United States and Northern Mexico, as well as some desert areas of Argentina (Benson and Darrow, 1981). In Mexico, the plant is located in the northern states (Chihuahua, Coahuila, Sonora, and northern Zacatecas) where it covers about 50 million hectares. It is traditionally used in Mexico to treat infections, kidney problems, gallstones, diabetes, rheumatism, arthritis, wounds, skin injuries, and dissolve tumors. It is used to treat tuberculosis, venereal diseases, cancer, and wounds, as well as an expectorant and tonic (Gonzalez, 1968). Also, it has been reported that *L. tridentata* possesses antibacterial lignans and flavonoids, cytotoxic lignans, cytotoxic triterpene glycosides, antioxidant lignans, anti-HIV lignans, furanoid lignans, and flavonoids.

Due to weather conditions where the gobernadora grows, this plant has high levels of molecules known as tannins or polyphenols. Polyphenols are weakly acidic nitrogenous organic compounds, amorphous, astringent taste, the most soluble in water, only a few in organic solvents; they are yellow, red, or brown and are located in the cytoplasm and cell vacuole as (Martins et al., 2010; Mueller-Harvey, 2001, Lekha and Lonsane, 1997;

Lewis and Starkey, 1968) secondary metabolites. These compounds have the property of tanning leather making them waterproof leathers which are resistant to attack by bacteria, heat, and abrasion (Swain, 1959). Also allow plants protect themselves from predators and microbial attack, which is possible thanks to its high-antioxidant power of these molecules because they have a toxic effect against potential pathogens resistance of certain tissues of the plant to attack viruses and microorganisms, partnering with the protein component of the virus or inactivating the microbial enzyme. Similarly where the tannins (quinone oxidation by enzyme phenolase) polymerize an insoluble barrier is formed, which prevents the microbial attack (Mendez, 1984).

Extraction of polyphenolic compounds (antioxidants) and antimicrobial compounds from a semi-this plant can be carried out by traditional extraction (Soxhlet, infusion, and maceration) and supercritical fluid extraction. These extraction methods can cause damage or loss of antioxidant activity in the compounds extracted by the use of high temperatures for long periods. That is why the interest in the application of sonochemistry for extraction of natural products has increased and is due to its advantages (Lianfu and Zelong, 2008; Rodrigues et al., 2007). This technique has proven to be one of the most important for the extraction of bioactive plant to be adaptable to a small or large scale, be cheap, uses less solvent, low temperatures, and time of final extraction, energy saving, it is also very useful for extracting heat labile and unstable compounds and therefore provides increased extraction yield (Liu et al., 2010; Ma et al., 2008; Liafun and Zelong, 2008).

The aim of this study was a comparative study of extraction reflux and assisted extraction ultrasound for polyphenolic compounds of gobernadora plant and, thus, to know which one of the two approaches is more effective for extracting polyphenolic compounds and measurement of its antioxidant potential.

8.2 EXPERIMENTAL

8.2.1 RAW MATERIAL

Larrea tridentata (Sesse & Mocino ex DC.) Coville plants were collected from the semidesert of Chihuahua (Northeast Coahuila, Mexico) during the summer of 2015. The plants were treated according to the method reported by Castillo et al. (2010) and Osorio et al. (2010). The dry material was ground and was stored in plastic dark containers at room temperature and protected from light to prevent oxidation of the compounds.

8.2.2 REACTIVE AND STANDARDS

The extractions mixtures were used distilled water and ethanol. The reagents used in this study were sodium carbonate, the Folin–Ciocalteu reagent (2N), 1,1-diphenyl-2-picrylhydrazyl (DPPH), gallic acid (GA), catechin (CAT), nordihydroguaiaretic acid (NDGA) purchased from Sigma-Aldrich (Saint Louis, MO). Solvents used were acetonitrile (ACN), acetic acid in methanol, and ethanol analytical grade.

8.2.3 REFLUX EXTRACTION METHODOLOGY

In the present study was selected to water and ethanol solvents because they are found by the Food and Drugs Administration for use in sanitizing products within approved food industries.

Reflux extraction was performed using 5 g of dried plant, was mixed with solvents (solid/liquid ratio 1 g/10 mL^{-1}) in Erlenmeyer flasks were covered with foil paper to prevent oxidation of compounds. Was used a water bath at 60±2°C using different concentrations of ethanol (25%, 50%, and 75% v/v) and extraction times were 1, 2, and 3 h.

8.2.4 ULTRASOUND EXTRACTION

The polyphenols extraction of *Larrea tridentata* by ultrasound methodology was done using an ultrasonic bath (BRANSON, model 2510) with a capacity of 10 L. Were used plastic containers with a capacity of 100 mL, which were covered with foil paper to prevent exposing the compounds to light. Solvent mixtures were the same as were used in the extraction by refluxing (25%, 50%, and 75% v/v) and sonication times were 20, 40, and 60 min (Muñiz et al., 2013). Ultrasound equipment was operated at a frequency of 40 kHz and at room temperature.

The plant extracts were filtered (Whatman no. 41) and were stored in amber containers until their analysis. The samples were diluted with water to determine total hydrolyzable polyphenols, condensed tannins, antioxidant activity, and a phytochemical profile by HPLC/MS. All measurements were performed in triplicate and under the same reaction conditions.

8.2.5 TOTAL PHENOLIC COMPOUNDS ASSAY

The total content of hydrolyzable polyphenols was analyzed by Folin–Ciocalteu assay modified in a microplate. In test tube was placed 20 µL of sample; the same tube was added 20 uL of Folin–Ciocalteu reagent, stirred, and allowed to stand for 5 min. After that, were added 20 µL of Na_2CO_3 (0.01 M) stirred and allowed to stand for 5 min. Then diluted with 125 µL of distilled water and read on a plate reader (Epoch) to 725 nm to determine hydrolyzable tannins. The response variable was the amount of total phenolic compounds in the extracts which was expressed as mg GA equivalents (GAE)/g of plant material using a regression equation and a GA calibration curve (R^2 0.995). The extract with the highest polyphenol content was selected for analysis by HPLC/MS.

8.2.6 CONDENSED TANNINS ASSAY

Condensed tannins were analyzed with the HCl-Butanol technique. After that, was added 3 mL of HCl/Butanol mix (1:9) and 0.1 mL of ferric reagent. After, all tubes were heated in a water bath at 100°C by 1 h. Then, were allowed to cool and the absorbance read in a UV/visible spectrophotometer (Varian 50 Bio spectrophotometer) at 460 nm. The samples concentration was obtained in CAT equivalents per mg of material per linear correlation according to the CAT standard curve of 0–1000 ppm.

8.2.7 ANTIOXIDANT ACTIVITY ASSAY

A 60 µM solution of DPPH in methanol was prepared and 193 µL of this solution was added to 7 µL of extract resuspend in distilled water (1:100) in a microplate. The mixture was shaken vigorously and allowed to stand at room temperature for 30 min. Then absorbance was measured at 517 nm (Epoch Gen 5 software). The capability to scavenge DPPH radical was calculated using the following equation:

8.2.8 HPLC/MS ASSAY

Phenolic profile of the extract was performed according to the antioxidant properties. The sample was previously filtered through a 0.22 μm nylon membranes (Millipore) and injected in the HPLC system (Varian Pro Star 330) under the following operation conditions: 5 μm column Optisil ODS (250 × 4.6 mm) and an injector 10 μL, 280 nm. The mobile phase was acetic acid and ACN (3%) with a flow rate of 10 μL min^{-1}. All standards were injected in the same way that the extract selected (Muñiz et al., 2013).

8.3 RESULTS AND DISCUSSIONS

The ANOVA detected an interaction between the extraction treatment (reflux/ultrasound) and the solvent (water/ethanol). Also, the interaction between the extraction treatment (reflux/ultrasound) and time (h/min) on the percentage of DPPH radical (Table 8.1), the concentration of total polyphenols hydrolyzable and condensed tannins.

TABLE 8.1 % DPPH Radical of Reflux and Ultrasound-assisted Extraction of *L. tridentata* Compounds.

Extraction	Time (h)	Solvent			
		% DPPH radical scavenging			
		H$_2$O	EtOH 25%	EtOH 50%	EtOH 75%
	0.33	61.7 ± 1.0	60.4 ± 5.0	59.6 ± 8.0	64.3 ± 1.9
Ultrasound	0.66	61.1 ±4.4	63.2 ± 1.5	58.5 ± 2.0	61.1 ± 3.6
	1	61.1 ± 1.1	61.7 ± 4.4	54.3 ± 6.1	50.5 ± 13.4
	1	57.8 ± 1.5	63.8 ± 1.5	64.5 ± 1.4	64.7 ±1.1
Reflux	2	62.6 ±3.4	66.1 ± 2.7	65.8 ± 1.9	69.0 ±1.1
	3	63.0 ±3.1	65.9 ± 3.9	67.3 ± 1.9	68.1 ± 1.3

Tukey test (0.05) showed a significant difference between treatment reflux and ultrasound-assisted extractions in the percentage of DPPH radical. The reflux-assisted extraction is the best treatment to increase DPPD radical. Furthermore, that nonsignificant difference was found in the % of DPPH in relation to the extraction time. Total hydrolyzable polyphenols were observed that as the concentration of ethanol as solvent extractant concentration increases mg of gallic acid equivalents (Fig. 8.1) is increased.

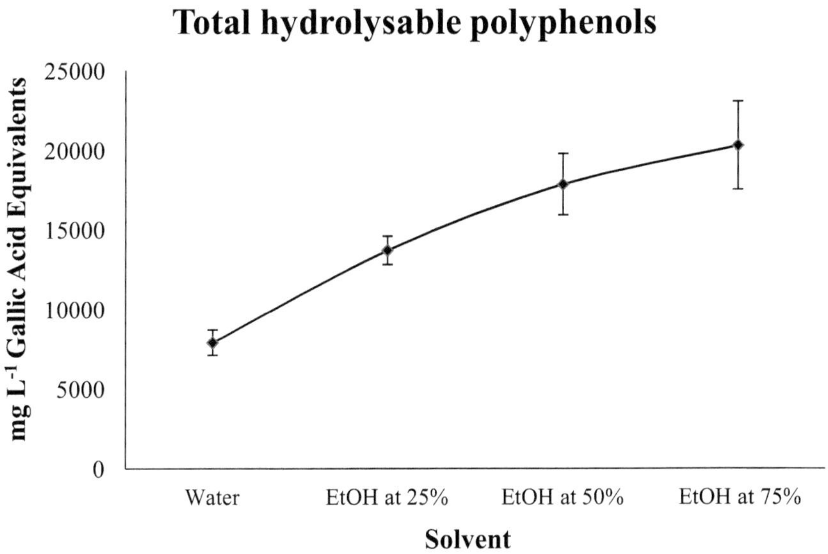

Total hydrolysable polyphenols

FIGURE 8.1 Graph of total polyphenol hydrolyzable by reflux extraction.

Condensed tannins showed an increase in the concentration of milligrams of CAT equivalents. Due to the increased ethanol content; however, condensed tannin concentration decreases to 75% ethanol (Fig. 8.2). This interaction was observed by Sultana et al. (2009), who investigated the effects of extracting solvents (absolute ethanol, absolute methanol, aqueous ethanol, and aqueous methanol) on the extraction of phenolic compounds from medicinal plants, obtaining the highest yields with alcohol/water mixtures.

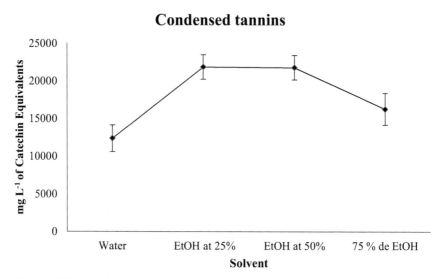

FIGURE 8.2 Graph of condensed tannins by reflux-assisted extraction.

Ultrasound-assisted extraction also showed to be the best for obtaining more gallic acid equivalents (total hydrolyzable polyphenols) and catechins (condensed tannins). This is attributed to the ultrasonic mechanism that is mainly the behavior of cavitation bubbles upon propagation of acoustic waves. The collapse of bubbles can produce chemical, physical, and mechanical effects (Muñiz et al., 2013; Wang et al., 2008), which result in disruption of biological cell walls, facilitating release of extractable compounds and enhancing mass transport of solvent from the continuous phase into plant cells (Muñiz et al., 2013; Ma et al., 2008).

The HPLC/MS analysis detected and confirmed the presence of NDGA (Marins et al., 2010), peak on the HPLC chromatogram showed the mass 302 (Fig. 8.3). However, this analysis also revealed other mass peaks that corresponded to other phenolic compounds of *Larrea tridentata* the table shows the mass and the structure of those compounds (Table 8.2).

Analysis and Quantification of Larrea tridentata Polyphenols

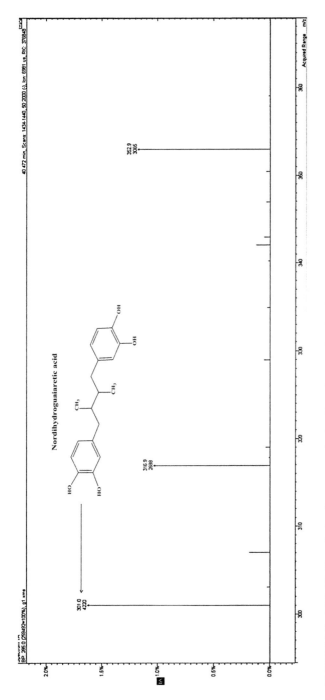

FIGURE 8.3 HPLC/MS analysis of an extract obtained by ultrasound-assisted extraction of *Larrea tridentata*.

106   Green Chemistry and Biodiversity: Principles, Techniques, and Correlations

TABLE 8.2   Characterized Compounds in the HPLC/MS.

m/z	Compound	Structure
192	Scopoletin or coumaric acid ethyl ester	
285	Luteolin-flavonoid	
286	Kaempferol-flavonol	
290	Epicatechin	
302	Nordihydroguaiaretic acid	
306	Gallocatechin	
318	*6,8-dihydroxy-kaempferol*	
348	Pentadecyl resorcinol	
366	Secoisolariciresinol-lignane	
880-881	Prodelphinidin trimmer	

8.4 CONCLUSIONS

The increase in the ethanol concentration is proportional to the increase of the hydrolyzable polyphenols and condensed tannins still the percentage of DPPH alone increased by 10% in the content of total hydrolyzable polyphenols and condensed tannins was increased with an increasing percentage of ethanol. This is because the solvent allows the extractions of compounds are not soluble in water. In the analysis of HPLC/MS allowed to identify 10 mass corresponding to compounds like coumaric acid, catechins, and lignans were identified. The most important compounds found nordihydroguaiaretic acid.

ACKNOWLEDGMENTS

The authors wish to express their gratitude to the National Council of Science and Technology (CONACYT, Mexico) for their financial support through the project PEI-222691. The authors also acknowledge Dr. Miguel Angel Medina and Alfredo Ivanoe Garcia for technical support.

KEYWORDS

- **antioxidants**
- ***Larrea tridentata***
- **polyphenols**
- **reflux extraction**
- **ultrasound-assisted extraction**

REFERENCES

Benson, L.; Darrow, R. A. *Trees and Shrubs of the Southwestern Deserts,* 3rd ed; University of Arizona Press: Tucson, 1981 Brinker, F. *Larrea tridentata* (D.C.) Coville (Chaparral or Creosote Bush). *Br. J. Phytother.* **1993,** *3,* 10–30.

Castillo, F.; Hernandez, D.; Gallegos, G.; Mendez, M.; Rodriguez, R.; Reyes, A.; Aguilar, C. N. In Vitro Antifungal Activity of Plant Extracts Obtained with Alternative Organic Solvents Against *Rhizoctonia solani* Kühn, *Ind. Crops Prod.* **2010,** *32,* 324–328.

González, M. H. Ecología y Distribución de la gobernadora. In Sumario de la Conferncia sobre Gobernadora (*Larrea tridentata*), Sul Ross State College: Alpine, TX. p 9.

108   Green Chemistry and Biodiversity: Principles, Techniques, and Correlations

Lekha, P. K.; Lonsane, B. K.Production and Applications of Tannin Acyl Hydrolase: State of the Art. *Adv.App. Microbiol.* **1997**, *44*, 215–260.

Lewis, J. A.; Starkey, R. L. Vegetable Tannins, their Decomposition and Effects on Decomposition of Some Organic Compounds. *Soil Sci.* **1968**, *106*, 241–247.

Lianfu, Z.; Zelong, L. Optimization and Comparison Ultrasound/Microwave Assisted Extraction (UMAE) and Ultrasonic Assisted Extraction (UAE) of Lycopene from Tomatoes. *Ultrason. Sonochem.* **2007**, *15*, 731–737.

Liu, Q. M.; Yang, X. M.; Zhang, L.; Majetich, G. Optimization of Ultrasonic-assisted Extraction of Chlorogenic Acid from *Folium eucommiae* and Evaluation of its Antioxidant Activity. *J. Med. Plants Res.* **2010**, *23*, 2503–2511.

Ma, Y. Q.; Ye, X. Q.; Fang, Z. X.; Chen, J. C.; Xu, G. H.; Liu, D. H. Phenolic Compounds and Antioxidant Activity of Extracts from Ultrasonic Treatment of *Satsuma mandarin* (*Citrus unshiu Marc.*) Peels. *J. Agri. Food Chem.* **2008**, *56*, 5682–5690.

Mabry, T. J.; DiFeo, D. R.; Sakakibara, M.; Bohnstedt, C. F.; Siegler, D. S. The Natural Products Chemistry of Larrea. In *Creosote Brush: Biology and Chemistry of Larrea in New World Deserts*; Hunziker, J. H. and DiFeo, D. F., Eds.; Hutchinson & Ross, Inc.: Dowden, 1977; pp 115–134.

Makkar, H. P. S.; Blummel, M.; Borowy, N. K.; Becker, K. Gravimetric Determination of Tannins and Their Correlations with Chemical and Protein Precipitation Methods. *J. Sci. Food Agri.* **1993**, *61*, 161–165.

Martins, S.; Aguilar, C. N.; De la Garza- Rodríguez, I.; Mussatto, S. I.; Teixeira, J. A. Kinetic Study of Nordihydroguaiaretic Acid Recovery from *Larrea tridentata* by Microwave-assisted Extraction. *J. Chem. Technol. Biotechnol.* **2010**, *85* (8), 1142–1147.

Molyneux, P. The use of the Stable Free Radical Diphenylpicryl-hydrazyl (DPPH) for Estimating Antioxidant Activity. *Songklanakarin J. Sci. Technol.* **2004**, *2*, 211–219.

Mueller-Harvey, I. Analysis of Hydrolysable Tannins. *Animal Feed Sci. Technol.* **2001**, *91*, 3–20.

Muñiz-Márquez, D. B.; Martínez-Ávila, G. C.; Wong-Paz, J. E.; Belmares-Cerda, R.; Rodríguez-Herrera, R.; Aguilar-González, C. N. Ultrasound-assisted Extraction of Phenolic Compounds from *Laurus nobilis* L. and their Antioxidant Activity. *Ultrason. Sonochem.* **2013**, *20*, 1149–1154.

Osorio, E.; Flores, M.; Hernandez, D.; Ventura, J.; Rodriguez, R.; Aguilar, C. N. Biological Efficiency of Polyphenolic Extract from Pecan Nuts Shell (Carya illinoensis), Pomegranate Husk (*Punica granatum*) and Creosote Bush Leaves (*Larrea tridentata Cov.*) Against Plant Pathogenic Fungi. *Ind. Crops Prod.* **2010**, *31*, 153–157.

Rodrigues, S.; Gustavo, A. S.; Pinto, F.; Fernandes, A. N. Optimization of Ultrasound Extraction of Phenolic Compounds from Coconut (*Cocos nucifera*) Shell Powder by Response Surface Methodology. *Ultrason. Sonochem.* **2007**, *15* (1), 95–100. DOI: 10.1016/j.ultsonch.2007.01.006

Sultana, B.; Anwar, F.; Ashraf, M. Effect of Extraction Solvent/Technique on the Antioxidant Activity of Selected Medicinal Plant Extracts. *Molecules* **2009**, *14*, 2167–2180.

Swain, T.; Hillis, E. The Phenolic Constituents of *Prunus domestica*. The Quantitative Analysis of Phenolic Constituents. *J. Sci. Food Agriculture* **1959**, *10*, 63–68.

Ventura-Sobrevilla, J. M. Tesis licenciatura. Biodegradacion de Taninos en Extractos de Gobernadora (*Larrea tridentata* Cov.) y Hojasen (*Fluorencia cernua D. C.*) mediante

Fermentación en Estado sólido usando *Aspergillus niger* PSH. Tesis Químico Farmaceutico-Biologo. Universidad Autónoma de Coahuila, 2006.

Wang, J.; Sun, B.; Cao, Y.; Tian, Y.; Li, X. Optimization of Ultrasound-assisted Extraction of Phenolic Compounds from Wheat Bran. *Food Chem.* **2008,** *106,* 804–810.

CHAPTER 9

Properties and Applications of the Phytochemical: Ellagic Acid (4,4',5,5',6,6'-Hexahydroxydiphenic Acid-2,6,2',6'-Dilactone)

RENÉ DÍAZ-HERRERA[1], PEDRO AGUILAR-ZARATE[2],
JUAN A. ASCACIO-VALDES[1], LEONARDO SEPÚLVEDA-TORRE[1],
JUAN BUENROSTRO-FIGUEROA[3], MONICA L. CHAVEZ-GONZALEZ[1],
JANETH VENTURA[1], and CRISTÓBAL N. AGUILAR[1*]

[1]*Bioprocesses and Bioproducts Group, DIA-UAdeC Food Research Department, School of Chemistry, Universidad Autónoma de Coahuila, Saltillo 25280, México*

[2]*Tecnológico Nacional de México, Instituto Tecnológico de Ciudad Valles, Ciudad Valles, SLP, México*

[3]*Research Center for Food and Development A.C., Delicias City, Chihuahua, México*

Corresponding author. E-mail: cristobal.aguilar@uadec.edu.mx

ABSTRACT

Ellagic acid (EA) is a polyphenolic compound derived from the hydrolysis of ellagitannins; it is a powerful bioactive component that is used in the industrial area due to their many properties and health benefits. In this review, we analyze and describe at detail the uses as bioactive in cosmetic, food and pharmaceutical industries, and the advantages of this polyphenol due to it is a compound within the reach of all that is found in nature in a lot of fruits, also we will mention the functions these compounds perform in the human body, as well a review of all the processes for the obtaining of this molecule, like alternative biotechnological processes, to improve and increase the production and recovery of the EA, making emphasis in solid state culture technique, and a list of the principal sources of EA, tannins, and their derivatives.

9.1 INTRODUCTION

Tannins are polyphenols, natural compounds with different molecular weights in the plant kingdom,[1] these compounds can be found in a large number of families of higher plants, with a molecular weight between 300 and 3000 Da, due to the chemistry of the tannins change according to their origin. Tannins are found in high concentrations in almost the entire plant, such as roots, bark, wood, leaves, fruits, and seeds.[2] Tannins are considered plants secondary metabolites,[3] they are compounds with low biodegradation and one of the most important natural antimicrobials in plants. Tannins have a specific function in plants, which is the defense system against animal attacks and the inhibition of growth of several micro-organisms such as bacteria, yeast, and fungi due to its astringent capacity[4] and the ability to form strong complexes with proteins and other macromolecules, due to these characteristics, it is said that tannins have bactericidal properties, however, there are micro-organisms capable of tolerating the presence of tannins and even using them as a carbon source.[5]

After cellulose, hemicellulose, and lignin, tannins are the fourth most abundant component and the most abundant phenolic polymers in plants. Tannins have important biological activities, this is why the extracts of plants rich in tannins have various uses (astringents, anti-inflammatories, antiseptics, and hemostatic pharmaceuticals).[6,7] When there are high concentrations of tannins in beverages (ice tea, beer, wine, fruit juices, and coffee flavor drinks) it results in the formation of precipitates due to the interaction of tannins with other molecules present in beverages. These effects of tannins can be reduced or eliminated through chemical or enzymatic treatments.[8]

Tannins are polyphenolic compounds, which according to their structure are divided into three groups; hydrolysable tannins which in turn are divided into two groups, since these have a sugar core linked by esterification to gallic acid (gallotannins) or hexahydroxydiphenic acid (HHDP) [ellagitannins (ETs)]; the condensed tannins formed by the monomer

Properties and Applications of Ellagic Acid

flavan-3-ol or flava-3,4-diol; and complex tannins that share the properties of the two previous groups of tannins, hydrolysable and condensed.[9] Of the four groups of tannins, ETs are the least studied group due to their diversity and chemical complexity.[4] When ETs are exposed to strong acidic or basic conditions, ester bonds are hydrolyzed and the HHDP group is released, which must go through a process of lactonization to become a more stable molecule known as EA (Fig. 9.1).[10,12]

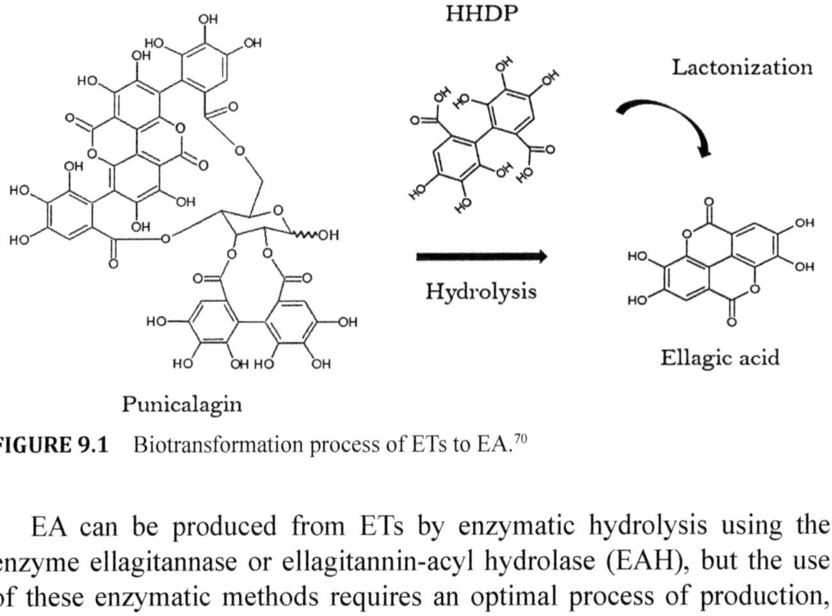

FIGURE 9.1   Biotransformation process of ETs to EA.[70]

EA can be produced from ETs by enzymatic hydrolysis using the enzyme ellagitannase or ellagitannin-acyl hydrolase (EAH), but the use of these enzymatic methods requires an optimal process of production, extraction, and purification of the enzyme. Ascacio–Valdés proposed the possibility of an alternative route for the biodegradation of ETs and AE production using pomegranate husk ETs as punicalagin and punicalin, which the micro-organism *Aspergillus niger* GH1 used as a carbon source for the production of AE by SSC.[13]

In addition, the EA has great importance due to numerous biological activities since it is believed that this compound works as an antimutagenic, antiviral, anticarcinogenic, antitumor agent with strong antioxidant activity.[14] Both ETs and the EA report a low absorption in the body because they are decomposed by the intestinal microbiota and because of the extensive metabolism a compound is generated that is better absorbed in the body known as urolithins. Since urolithins are catabolites derived from ETs that can be absorbed and can reach different tissues in the body, there are no studies about the pharmacokinetics and tissue distribution of urolithin after direct ingestion in humans. However, there is indirect evidence of these two aspects after the intake of foods rich in ETs and EA, this is mainly the reason why it is thought that urolithins are the molecules responsible for the biological effects observed as a consequence of consuming pomegranate or other fruits rich in ETs.[15]

Due to the importance of tannins and EA in industrial areas such as pharmaceutical, cosmetic, and food, the objective of this article is to review the generalities and properties of the EA, since lately the demand for this compound has been increasing due to its possible health benefits, as well as the production methods, that have been improving over the years and for the correct use of this biocomposite it should be considered as important factors the solubility and bioavailability, which can directly affect the properties of this bioactive component and for which it is a molecule of great importance.

9.2   ELLAGITANNINS

The ETs are part of the group of hydrolyzable tannins, they are polymers of HHDP and a molecule of glucose or quinic acid,[16] they are amorphous and with an astringent taste, they act as weak acids and are considered secondary metabolites of plants since they are not related directly in the primary metabolism, they are found in the cytoplasm and cellular vacuoles.[11] It should be mentioned that there are also ETs that cannot be hydrolyzed, this is due to an additional coupling in the carbon–carbon bonds in the polyphenolic residues with the polyol group present in the molecule, however, they are classified as hydrolyzable tannins, such as vescalagin.[17] ETs are the most abundant phenolic compounds in plants among hydrolyzable tannins.[18] The main sources of these compounds are fruits, such as pomegranate, strawberry, raspberry, blackberry, and grapes.[16] The nature of the bonds between the monomers determines the bases for these molecules classification.[19] The ETs by their properties are considered high value in the human diet since it is known that they participate in the prevention of degenerative diseases, such as cancer and cardiovascular diseases because they show antioxidant properties, which is important for the absorption and neutralization of free radicals in the body.

Properties and Applications of Ellagic Acid

This activity provides ETs with great phytochemical relevance because in addition of being considered natural antioxidants, ETs have other important properties for human health,[11] such as, anticancer, antiatherosclerotic, anti-inflammatory, antihepatotoxic, antibacterial, and repressor of HIV replication.[17] In recent years, the obtaining and biodegradation of these compounds has been the focus of research and several studies due to their commercial and scientific importance.[4]

9.3 ELLAGIC ACID

EA (4,4',5,5',6,6'-hexahydroxydiphenic acid-2,6,2',6'-dilactone), a lactone that is derived from gallic acid molecules and known to be the residue of the hydrolysis of ETs (Fig. 9.2),[20] is a hydroxybenzoic acid that can be present in its free form in some plant species, such as fruits and nuts as a product their metabolism, among these are raspberries, currants, grapes, and strawberries. Commonly, this compound is found in the vacuoles of plants as hydrolyzable and water-soluble ETs.[20,22]

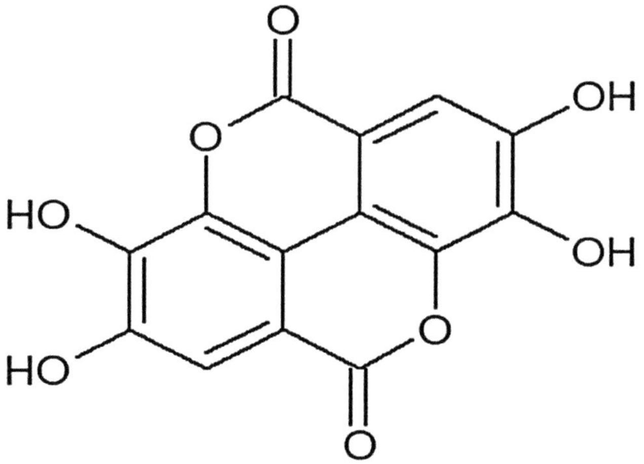

FIGURE 9.2 EA molecule.

In recent years, EA has aroused commercial interest as it is a high-value compound and has been used in the industrial sector, such as food, cosmetics, and pharmaceutical because it can be used for its properties as an antioxidant, antitumor, antiviral, antimicrobial, and anticancer.[5,13] However, due to its high production cost, all the applications of this compound have not yet been exploited.[14]

9.4 NATURAL SOURCES

The natural compounds that are derived from plants are part of our daily diet, a diet rich in fruits, vegetables, and cereals is associated with a decrease in the risk of acquiring chronic degenerative diseases,[24] so in recent years, leading a healthier lifestyle has become increasingly important and this increases the need to find new natural products. Among these compounds, we can find the EA.[4]

EA is present in a wide variety of plants, fruits, and nuts in nature.[20] One of the main sources of EA is the vegetable source and this compound can be found almost anywhere in the plant mainly in leaves, branches, roots, stems, fruits, and bark,[25] Table 9.1 shows some of the main sources of EA. It is known of the existence of EA in several bushes of berries, such as raspberry, blueberry, blackberry, in some seeds, and nuts and fruits, such as pomegranate.[4,18,24,26]

There is evidence of the presence of EA in the stems of candelilla, which has added great relevance to this plant species, now as a good source of EA,[20] candelilla (*Euphorbia antisyphilitica* Zucc.) is a plant with great economic relevance in the north of Mexico, one of its main uses is the production of candelilla wax from the stem, this wax has several applications, however, its extraction generates about 140 tons of waste per year[4]; although there is no information or previous studies on the recovery of this compound, which could be used by several industrial sectors.[120]

TABLE 9.1   EA Sources.

Scientific name	Common name	EA (mg g^{-1})
Castanea sativa	Chestnut	89.0
Juglans regia	American walnut	28.9
Quercus robus y alba	Oak tree	19.0–63.0
Rubus occidentalis	Blackberry	1.5–2.0
Euphorbia antisyphilitica	Candelilla	7.9
Flourensia cernua	Tarbush	1.59

Properties and Applications of Ellagic Acid

TABLE 9.1 *(Continued)*

Scientific name	Common name	EA (mg g^{-1})
Rubus idaeus	Blueberry	1.2–1.5
Turnera diffusa Willd	Damiana	0.87
Jatropha dioica	Sangre de Drago	0.81
Punica granatum	Pomegranate	0.78
Bertholletia excelsa	Brazil nut	0.59
Fragaria ananassa	Strawberry	0.4–0.6
Carya illinoinensis	Pecan nut	0.33

Source: Refs. [4, 20, 21, 25].

9.5 A BIOPROCESS TO PRODUCE EA

Attempts have been made to obtain EA using acid addition technique; however, due to nonspecific reactions, the desired product is not obtained, which results in similar structures and varied by-products. Because of this, the interest in biological methods increased. In order for the reaction to be specific and EA can be produced from ETs, the synergistic action of an enzyme capable of breaking the chemical bonds in the monomeric ETs is required. For this reaction enzymes are used to break the ester bonds in the molecule, These enzymes are produced mainly by micro-organisms for the formation of other compounds such as EA.[14] For this, strategies, such as SSC, have been applied to induce the micro-organisms to produce these enzymes.

SSC is a traditional method that has been used for centuries for the production of traditional foods in eastern countries.[28] Culture in the solid medium is defined as the growth of micro-organisms using solids as support in the absence or near absence of free water[29]; however, the substrate must contain a sufficient amount of moisture to cover the metabolic demands of the micro-organism and this can grow correctly.[30] In this type of fermentation, the amount of water in the system must not exceed the saturation capacity of the support where the micro-organism will grow.[31]

A great advantage in this type of fermentation is a large number of solid materials that can be used as support, which are classified into two categories: noninert materials and inert materials. In the first category, a distributed and humidified solid (grain of cereal, flour, bran, sawdust, etc.) fulfills functions both as a support and as a source of nutrients for the micro-organism, unlike the inert solids (polyurethane foam), which only fulfill the function of support, soaked with a suitable culture medium. Another advantage that attracts attention is the use of agro-industrial waste as noninert supports since it makes this process quite economical.[32]

SSC provides the opportunity for the utilization of agro-industrial wastes that usually end up accumulating garbage, besides that this process becomes less expensive because it requires less energy and is friendly to the environment since it produces less wastewater than the submerged culture.[30] SSC is innovative due to new applications that the industry sector has found in recent years, this technique has a great future, in particular by the valorization that is given to the agro-industrial byproducts, the utilization and biodegradation of solid wastes, bioremediation of organic pollutants in soils, and the fact that it reduces atmospheric pollutants by biofiltration.[33]

SSC over time has been able to change and continues to develop, there are studies comparing submerged culture with SSC, which demand higher yields, a higher concentration of enzyme with a greater potential for the recovery of this and other advantages of SSC compared to submerged culture.[34]

Saavedra et al. (2005) reported that the production of EA has not been studied due to the high price in production and the large number of byproducts that are generated after the biodegradation of ETs, since this affects the purification and recovery of EA. Until now, the study for the biodegradation of ETs using biological methods, whether enzymatic or microbial, turns out to be a very promising topic. In 2007, the production of EA was reported by microbial culture in SSC using polyurethane foam with an aqueous extract obtained from pomegranate peel (*Punica granatum*) as a carbon source.[35]

The authors of this work attribute the biodegradation of the molecule to a new enzyme different from tanin-acyl hydrolase, in addition to showing that it is possible to produce EA using a biotechnological system.[25] Table 9.2 shows some previous works in the production of EA by the technique of SSC.

TABLE 9.2 Production of EA by SSC.

Micro-organisms	Substrate	Recovered EA	Source
A. niger GH1	Pomegranate husk ETs	42.02 mg/g	[11]
A. niger GH1	Creosote leaves ETs	138.6 mg/g	[71]
A. niger GH1	Pomegranate ETs	175 mg/g	[13]
A. niger GH1 & PSH	Pomegranate husk ETs	6.3–4.5 mg/g	[9]
A. niger GH1	Pomegranate husk ETs	132.62 mg/g	[72]
Lentinus edodes	Blueberry pomace	260–350 mg/g	[27]
A. niger PSH	Pomegranate husk ETs	330.89 mg/g	[73]

Tanin-acyl hydrolase (tannase) is an enzyme that catalyzes the breakdown of hydrolyzable tannins or gallic acid esters. This enzyme can be produced in the presence of tannic acid by some micro-organisms, such as bacteria, yeast, and fungi, the latter being the largest producers of this enzyme. Tannase is used by different industrial sectors such as food, beverages, brewer, pharmaceutical, and chemical.[36] The search for new types of tannases with different properties that are as useful for industries and have new applications has been a frequent work since this enzyme was discovered more than 100 years ago.[37]

Recently Ascacio-Valdés et al. (2013) reported that the strain of *Aspergillus niger* GH1 can grow on SSC producing an enzyme called EAH.[11] The EAH is the enzyme responsible to produce EA since it acts in the biodegradation of ETs. It is believed that this newly discovered enzyme is responsible for catalyzing the hydrolysis of the HHDP group of ETs for the formation of EA. Therefore, it is necessary to continue with new investigations to understand with clear information the route of biodegradation of ETs.[39]

9.6 PROPERTIES AND APPLICATION OF EA

Because EA has important chemical properties, this has been reported for its positive physiological effects on the health of those who consume it, therefore, it is important to mention all the applications of this compound.[25] This is the reason why the pharmacological potential of EA has allowed it to be considered as one of the most important bioactive components in recent years.[40]

Among these properties, there are treatments that show detention in the cell cycle along with the induction of programmed cell death (apoptosis) and the inhibition of tumor formation in vivo.[22] Some cases have been reported in the decrease of cancer cells by ETs and EA even after the symptoms appear. In addition, according to studies, indicate that the EA can prevent the development of cells infected by the human papillomavirus,[4] as well as control hemorrhages in both humans and animals, this is due to its ability to activate the Hagerman factor,[9] among others. Studies conducted with animals showed that for the compound to influence the human body, the minimum recommended amount of EA intake for a 65 kg adult is 260 mg per day.[19]

9.6.1 ANTIMUTAGENIC AND ANTICARCINOGENIC ACTIVITY

Cancer is one of the leading causes of death in developed countries, only after heart diseases. The reason why now exists many types of cancer is largely attributed to chemical agents, for example, bladder cancer, lung, and hepatocellular carcinomas.[41] AE in addition of being a chemopreventive compound[42] it has been shown to regulate carcinogenesis by inhibiting enzymes, this in both in vivo and in vitro studies in rodents, preventing the mutation of healthy cells; however, the mechanism of action is not entirely clear.[41,43]

It has been reported that EA interferes with some diseases dependent on angiogenesis. The process of antiangiogenesis against VEGFR-2 has been considered as an important strategy as a therapy against breast cancer since it is known that the EA inhibits a series of VEGF factor that induces angiogenesis processes, such as proliferation, migration, and vessel formation in endothelial cells.[44]

As the understanding of the cancer process has increased, preventive methods for this disease have also increased, which is expected that due to these preventive methods and changes in diet can change the incidence of this disease, for the time being another alternative is the chemical intervention, which offers attractive results in a short time.[43]

9.6.2 ANTIOXIDANT ACTIVITY

Antioxidants are chemical compounds that in low concentrations prevent or delay the oxidation of various substances by not allowing their molecules

Properties and Applications of Ellagic Acid 121

to bind to oxygen,[45,46] this process occurs both in food and human body; in foods the antioxidants help to prevent physiological alterations, such as oxidative stress and the formation of free radicals. It has been scientifically documented in many cases that antioxidants are health-enhancing and therefore they play an important role in the prevention of chronic and noncommunicable diseases, hence the importance of eating foods with a high content of antioxidants, such as fruits and vegetables.[46]

The phenolic compounds have antioxidant properties since they are known to eliminate free radicals; therefore, in recent years, there has been an increase in the interest to find new sources for obtaining polyphenolic compounds, and it is due to their antioxidant properties. Red fruit juices, such as berries or grape juice have received a lot of attention, however, a full investigation on the subject has not yet been carried out.[47,48]

EA is also used as a powerful antioxidant in different industrial sectors.[20] Its antioxidant capacity is believed to be due to the four hydroxyl groups in the structure, which interact and scavenge hydroxyl and superoxide anions radicals. Experiments in vivo indicate that AE has a high activity against reactive physiological species, therefore, the use of this compound is a natural way to eliminate free radicals from the human body.[25,49]

9.6.3 ANTIMICROBIAL ACTIVITY

The antimicrobial activity of phenolic compounds present in plants has been a very studied subject, since they can control pests and inhibit the growth of micro-organisms in the plant, it has been investigated the activity they can have against pathogens in the human body to invest in the development of new and healthy food ingredients and medicines that can provide these antimicrobial properties. Berry fruits and their extracts have been used as natural antimicrobials since it is known extracts can inhibit the growth of pathogens in the gastrointestinal tract, thus, preventing gastrointestinal and urinary tract diseases. It has been reported that blueberries can control and suppress the growth of micro-organisms, such as *Helicobacter pylori* and *Listeria monocytogenes*.[50]Although there are several reports in the literature on the antimicrobial activity of some fruits, pomegranate being one of them, no article relates this activity to its chemical composition.[51] Machado et al. (2002) identified pomegranate tannins as the compounds responsible against the activity of 18 strains of

Staphylococcus aureus, punicalin α and β and punicalagin α and β were identified, with the greatest responsibility being punicalagin anomers.[25]

Pomegranate punicalagin (*Punica granatum*) showed the power to inhibit strains of *Escherichia coli*, *Pseudomonas aruginosa*, *Cryptococcus neoformans*, and *Staphylococcus aureus* resistant to methylcyanin,[52] it is believed perhaps by the astringent properties and toxicity of the tannins. EA has also been shown to be a potent antimicrobial against pathogens, such as *Staphylococcus aureus* by inhibiting its growth.[53] Whose antimicrobial potential is similar to ETs.[52]

9.6.1 ANTIVIRAL ACTIVITY

Extracts of plants have been used for hundreds of years for the treatment of diseases, to this day. Mexico is the fifth country with a greater variety of plants in the world, counting with almost 3500 species of medicinal plants with biological potential. There have been considerable scientific studies with herbs to isolate compounds with potential medical benefits. Recently, interest in antiviral properties in particular activity against HIV-1 has been increasing.[54,55] They are looking for new alternatives for treatment against infection of HIV, under the theme that are safer and cheaper, since it is a matter of great global demand.[54]

EA and ETs are studied extensively due to their ability to inhibit the replication of the human immunodeficiency virus (HIV),[25] anti-HIV activity has been related to the HHDP groups of ETs, which inhibit the reverse transcriptase enzyme, the expression of HIV antigens in humans and the in vitro replication of the virus when they are present during cell infection.[56,57]

These activities are believed to depend on the number of HHDP groups and the cytotoxicity of the molecule, some authors suggest that the antiviral activity of the tannins against the herpes virus is because they inhibit the absorption of the virus, joining the components of the viral envelope. There are reports on the activity of various hydrolyzable tannins that have been analyzed for activity against the herpes virus,[58] these activities are believed to depend on the number of HHDP groups and the cytotoxicity of the molecule, some authors suggest that the antiviral activity of the tannins against the herpes virus is because they inhibit the absorption of the virus, joining the components of the viral envelope. These compounds

Properties and Applications of Ellagic Acid

have also been tested against other viruses such as influenza virus type A, parainfluenza virus, and hepatitis type A.[57] Unquestionably, the antiviral activity against herpes simplex virus type 1 and 2 and HIV-1 constitute one of the most important reports about the biological activity of these compounds associated with their chemical structure.[56]

9.7 INDUSTRIAL USES

The properties of EA are used in the food industry for the preparation of nutraceutical beverages, food supplements, and the conservation of perishable foods by using it as an inhibitor of micro-organisms. This compound, together with candelilla wax, is used for the preparation of edible coatings for fruits, such as avocados, to increase their shelf life and protect it from pathogens.[59,60]

Within the beverage industry, AE is used for the clarification of beverages, since they precipitate substances, such as carbohydrates and proteins avoiding the formation of colloids, mainly in the brewing industry,[25] it is also used to avoid the formation of creams during the elaboration of Instant tea through a de-esterification process.[37] EA is also used in the wastewater treatment industry since it can precipitate and reduce metal ions in an aqueous solution,[61] and the manufacture of skin lightening lotions, such as AEA, reduce the generation of melanin.[25]

Also compounds, such as ETs and enzymes derived from tannase are used in the wine industry, ETs protect the other components present against oxidation and the activity of the enzyme helps to maintain a high content of aromatic compounds, an appropriate color, and avoid the formation of turbidity in the wine.[62,63]

9.8 BIOAVAILABILITY

The issue of the bioavailability of a bioactive component turns out to be a very important factor that can affect the biological activities of a functional or nutraceutical food, however, this issue has not received the attention it deserves.[64,65] Bioavailability is defined by the FDA as "the rate and degree in which the therapeutic portion of a molecule is absorbed and becomes available at the site of action of the drug." The bioavailability in general is the product of (1) the amount of lipophilic compound that becomes accessible for intestinal absorption, (2) the fraction of the compound that can be transported through the epithelium of the small intestine, and (3) the fraction of the absorbed compound that reaches the systemic circulation without being metabolized.

The effectiveness of nutraceuticals to provide therapeutic benefits depends on the bioavailability of the active ingredient in the body, for the compound to be absorbed, this must be solubilized or dispersed in the intestinal lumen, after absorption in the intestine, compounds are submitted to an extensive metabolism, which could change its chemical structure and could alter its functionality.[66] Many compounds have low aqueous solubility and this is associated with low oral bioavailability, there are other factors such as stability against gastric pH, metabolism by intestinal flora,[64] low transport coefficient through the intestinal mucosa, instability against physiological changes, susceptibility to rapid metabolic transformation, etc. They can significantly reduce the effectiveness of nutraceuticals in the prevention of diseases since the bioavailability of nutraceuticals can be defined as the fraction of a dose administered orally that eventually reaches the systemic circulation.[66]

EA is a polyphenol that must be released from nonbioavailable ETs of pomegranate, nuts, and strawberries to be absorbed. The systemic effects of EA may be hampered by its limited bioavailability, which may depend on its low solubility in aqueous media (especially at acid pH, where most of this compound is not ionized), in vivo hydrolysis of punicalagin to release EA, the union of AE and/or punicalagin to proteins that can diminish its accessibility, the transfer, and absorption in the intestine and the catabolism of the intestinal microbiota to turn the EA into urolithins. This complex sum of factors may result in limited bioavailability of EA.[65]

The poor absorption of EA, according to reports, affects its antimutagenic activity in vivo since the concentration levels of the compound are very low in plasma and/or in target cells after oral administration, although these low levels in plasma can also be attributed due to its extensive metabolic transformation and degradation prior to being absorbed.[67] A higher intake of free EA does not mean that this improves the bioavailability this compound, but the biological action in the gastrointestinal tract may be greater and may promote the production of urolithins, which may contribute to the final benefits.[65]

Absorption, bioavailability, and pharmacokinetics of both the ETs and EA, administered orally, have not been adequately investigated. Therefore,

Properties and Applications of Ellagic Acid

it is necessary to perform these studies to determine the fate of these bioactive compounds, which are widely used as botanical ingredients and supplements.[67] A limited number of pharmacokinetic studies in humans have reported the bioavailability of EA on the intake of lyophilized pomegranate and raspberry juices or extracts, therefore, current knowledge about the bioavailability of EA and ETs has been reserved only for animal studies.[65,67]

The first study to evaluate the bioavailability of pomegranate husk ETs was carried out in rats, and the result after the ingestion of large amounts of pomegranate husk ETs showed that the main metabolites detected in plasma and urine were urolithins and low amounts of EA. Interestingly, a small amount of punicalagin was also detected, particularly after a long time of consumption; however, subchronic oral administration of large amounts of pomegranate ETs was not toxic in rats, where these compounds were found to be metabolized to urolithins. Urolithins are derivatives from dibenzofuran-6-one with different hydroxyl substitutions. Chemically they can be considered a combination between coumarin and isocoumarin (benzocoumarin). Urolithins constitute a complete family of metabolites produced by the decarboxylation of one of the lactones in the AE (lactonase and decarboxylase activity) and the elimination of the hydroxyl groups (dehydroxylase activity). After decarboxylation, the first metabolite produced is urolithin M5 and from this other urolithin, isomers are produced by the elimination of hydroxyl groups in different positions (Fig. 9.3). The name urolithin was given to a pair of metabolites isolated from kidney stones of sheep that were called urolithin A and B, urolithins are not common molecules in nature, but these molecules have been reported in plants rich in ETs, such as the leaves of pomegranate tree.[15]

The urolithins have been proposed as microbial metabolites that are biosynthesized in the colon and can circulate in plasma at 24 h after ingestion. In one study, EA and its metabolites (urolithins) were found in plasma after consumption of pomegranate extract, the presence of urolithins in plasma after 8 h is believed to be attributed to the action of the gut microbiota on previously consumed ETs or due to the enzymatic action of the plasma on EA present in the circulation.[68]

Urolithins circulate in plasma as glucuronides and conjugated sulfates in concentrations between 0.2 and 20 µM; therefore, it is correct to think that the beneficial health effects of products rich in ETs and EA are

126 Green Chemistry and Biodiversity: Principles, Techniques, and Correlations

associated with this molecule produced in the intestine, this is the reason why it is essential to evaluate the biological effects of this metabolite. Recent in vitro research has shown preliminary evidence about the anti-inflammatory, anticarcinogenic effects (colon and prostate cancer), antioxidant, antimicrobial effects, and cardiovascular protection of urolithins, supporting their potential contribution to health, therefore, the transformation from EA to urolithins plays a critical role in the biological activities attributed to these compounds.[15,69]

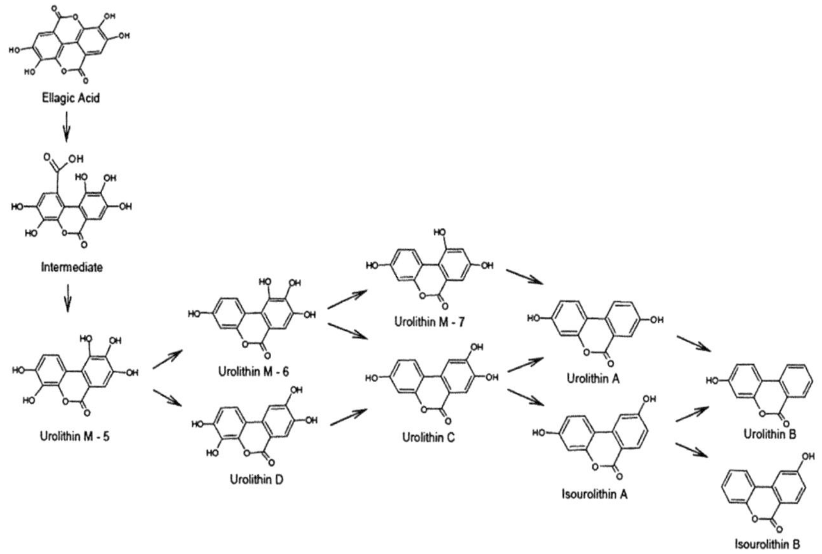

FIGURE 9.3 Urolithins production from EA.
Source: Reprinted with permission from Ref. [15]. https://www.hindawi.com/journals/ecam/2013/270418/

The number of in vivo studies is quite limited, but their results show preventive effects of urolithins on the intestine, which encourages further in vivo studies and studies on the mechanism as they are necessary to clarify the effects of urolithins. The production of urolithins from ETs has been reported in different animals that feed on bark, wood, and oak leaves as ruminants, beavers, and squirrels, these animals produce urolithins in their intestines and urolithins are found in their feces and plasma.

An Iberian pig was used as a model for the study of the production of urolithins from ETs. This was fed with oak acorns, which is a food rich in ETs, to evaluate the metabolism and tissue distribution of these tannins and understand their metabolism in humans. This study showed that different urolithins were produced in the intestine starting with tetrahydroxy-urolithin after the elimination of one of the lactones of EA followed by the elimination of the hydroxyl groups and result in urolithins A and B.

In this study it was found that urolithin A was produced from EA, punicalagin and walnut-rich ETs extract by the fecal microbiota of six volunteers, demonstrating for the first time the production of these metabolites by the human intestinal flora. Because there was a lot of variability in ex vivo results in relation to the in vivo results, it is thought that the differences in the composition of the gut microbiota of each person affects the production of urolithin and, therefore, affects the possible beneficial effects after consuming the food.[15]

A study was carried out on the main groups of bacteria present in fecal samples of healthy volunteers and their ability to produce different urolithins after being fed with nuts and pomegranate extracts, where it was discovered that the urolithin producing strain could be of the genus *Gordonibacter. G. urolithinfaciens*. The analysis of fecal samples of the volunteers showed the presence of urolithin A, B, isourolithin A, and minor metabolites such as urolithin-M5, M6, M7, C, and D, these results coincide with the results obtained with urine sample of the volunteers. In this study, they were able to confirm the three phenotypes for urolithin after ingesting walnuts or pomegranate extract,[69] according to the type and amount of urolithins produced by each individual. Phenotype 0 (does not produce urolithins), phenotype A (produces urolithin A) and phenotype B (produces urolithin A, isourolithin A, and B and urolithin B), however, the impact of catabolism of EA in a pharmacokinetic study has not been explored.[65]

More research will be needed on the *Gordonibacter* strain, to know if the population of these bacteria can be increased and how this could improve the metabolism of AE in vivo. Correct identification and characterization of these bacteria could be the key to develop new prebiotics with health benefits.[69]

9.9 FINAL REMARKS

The great importance of compound EA results in the research for new sources and new methods for obtaining this metabolite. In recent years, biotechnological processes have become increasingly important as they offer many advantages in cost, environmental and recovery over chemical methods. A big problem is the low bioavailability of EA, since this factor can affect the properties that give the name of bioactive molecule, due to this, studies were conducted in animals and humans, in which it was possible to find a molecule with higher solubility derived from the metabolism of EA in the body, urolithins are bioactive metabolites and it is believed that these are responsible for the activities attributed to the EA. Therefore, there is still a lot to be investigated regarding these two functional molecules and to obtain better results it would be necessary to carry out more studies in humans.

ACKNOWLEDGMENT

The authors thank CONACYT (National Council of Science and Technology, Mexico) for the support provided for this work.

KEYWORDS

- bioactive compounds
- ellagitannins
- polyphenols
- solid-state culture
- urolithins

REFERENCES

1. Herderich, M. J.; Smith, P. A. Analysis of Grape and Wine Tannins: Methods, Applications and Challenges. *Aust. J. Grape Wine Res.* **2005,** *11* (2), 205–214.
2. Mingshu, L.; Kai, Y.; Qiang, H.; Dongying, J. Biodegradation of Gallotannins and Ellagitannins *J. Basic Microbiol.* **2006,** *46* (1), 68–84.

Properties and Applications of Ellagic Acid

3. Aguilar, C. N.; Rodríguez, R.; Gutiírrez-Sánchez, G.; Augur, C.; Favela-Torres, E.; Prado-Barragán, L. A.; Ramírez-Coronel, A.; Contreras-Esquivel, J. C. Microbial Tannases: Advances and Perspectives. *Appl. Microbiol. Biotechnol.* **2007**, *76* (1), 47–59.

4. Aguilera-Carbo, A.; Augur, C.; Prado-Barragán, L. A.; Favela-Torres, E.; Aguilar, C. N. Microbial Production of Ellagic Acid and Biodegradation of Ellagitannins. *Appl. Microbiol. Biotechnol.* **2008**, *78* (2), 189–199.

5. Sepúlveda, L.; Ascacio, A.; Rodríguez-Herrera, R.; Aguilera-Carbó, A.; Aguilar, C. N. Ellagic Acid: Biological Properties and Biotechnological Development for Production Processes. *Afr. J. Biotechnol.* **2011**, *10* (22), 4518–4523.

6. Chávez-González, M.; Rodríguez-Durán, L. V.; Balagurusamy, N.; Prado-Barragán, A.; Rodríguez, R.; Contreras, J. C.; Aguilar, C. N. Biotechnological Advances and Challenges of Tannase: An Overview. *Food Bioprocess Technol.* **2012**, *5* (2), 445–459.

7. Kraus, T.; Dahlgren, R.; Zasoski, R. Tannins in Nutrient Dynamics of Forest Ecosystems: A Review. *Plant Soil* **2003**, *256*, 41–66.

8. Rodríguez-Durán, L. V.; Valdivia-Urdiales, B.; Contreras-Esquivel, J. C.; Rodríguez-Herrera, R.; Aguilar C. N. Novel Strategies for Upstream and Downstream Processing of Tannin Acyl Hydrolase. *Enzyme Res.* **2011**, *2011* (1), 823619.

9. Robledo, A.; Aguilera-Carbó, A.; Rodriguez, R.; Martinez, J. L.; Garza, Y.; Aguilar, C. N. Ellagic Acid Production by *Aspergillus niger* in Solid State Fermentation of Pomegranate Residues. *J. Ind. Microbiol. Biotechnol.* **2001**, *35* (6), 507–513.

10. Khanbabaee, K.; van Ree, T. Tannins: Classification and Definition *Nat. Prod. Rep.* **2001**, *18* (6), 641–649.

11. Ascacio-Valdés, J. A.; Buenrostro, J. J.; De la Cruz, R.; Sepúlveda, L.; Aguilera, A. F.; Prado, A.; Contreras, J. C.; Rodríguez, R.; Aguilar, C. N. Fungal Biodegradation of Pomegranate Ellagitannins. *J. Basic Microbiol.* **2013**, *54* (1), 28–34.

12. Hussein-Elgailani, I.; Yacoub, C. Methods for Extraction and Characterization of Tannins from Some Acacia Species of Sudan. *Pakistan J. Anal. Environ. Chem.* **2016**, *17* (1), 43–49.

13. Buenrostro-Figueroa, J.; Huerta-Ochoa, S.; Prado-Barragán, A.; Ascacio-Valdés, J.; Sepúlveda, L.; Rodríguez, R.; Aguilera-Carbó, A. Aguilar, C. N.; Continuous Production of Ellagic Acid in a Packed-bed Reactor. *Process Biochem.* **2014**, *49* (10), 1595–1600.

14. Huang, W.; Niu, H.; Li, Z.; He, Y.; Gong, W.; Gong, G.; Optimization of Ellagic Acid Production from Ellagitannins by Co-culture and Correlation Between Its Yield And Activities of Relevant Enzymes. *Bioresour. Technol.* **2008**, *99* (4), 769–775.

15. Espín, J. C.; Larrosa, M.; García-Conesa, M.; Tomás-Barberán, F.; Biological Significance of the gut Microbial Ellagic Acid-derived Metabolites Urolithins. *Evidence-Based Complement. Altern. Med.* **2013**, *2013*, 1–15.

16. Kool, M. M.; Comeskey, D. J.; Cooney, J. M.; McGhie, T. K.; Structural Identification of the Main Ellagitannins of a Boysenberry (Rubus loganbaccus ?? baileyanus Britt.) extract by LC-ESI-MS/MS, MALDI-TOF-MS and NMR Spectroscopy. *Food Chem.* **2010**, *119* (4), 1535–1543.

17. Landete, J. M. Ellagitannins, Ellagic Acid and their Derived Metabolites: A Review About Source, Metabolism, Functions and Health. *Food Res. Int.* **2011**, *44* (5), 1150–1160.
18. Karonen, M.; Parker, J.; Agrawal, A.; Salminen, J. P. First Evidence of Hexameric and Heptameric Ellagitannins in Plants Detected by Liquid Chromatography/ Electrospray Ionisation Mass Spectrometry. *Rapid Commun. Mass Spectrom.* **2010**, *24* (24), 3151–3156.
19. Clifford, M. N.; Scalbert, A. Review Ellagitannins: Nature, Occurrence and Dietary Burden. *J. Sci. Food Agric.* **2000**, *80*, 1118–1125.
20. Ascacio-Valdés, J. A.; Aguilera-Carbó, A.; Martínez-Hernández, J. L.; Rodríguez-Herrera, R.; Aguilar, C. N. *Euphorbia antisyphilitica* Residues as a New Source of Ellagic Acid. *Chem. Pap.* **2010**, *64* (4), 528–532.
21. Häkkinen, S. H.; Kärenlampi, S. O.; Mykkänen, H. M.; Heinonen, I. M.; Törrönen, a. R. Ellagic Acid Content in Berries: Influence of Domestic Processing and Storage. *Eur. Food Res. Technol.* **2000**, *212*, 75–80.
22. Suzuki, N.; Masamune, A.; Kikuta, K.; Watanabe, T.; Satoh, K.; Shimosegawa, T. Ellagic Acid Inhibits Pancreatic Fibrosis in Male Wistar Bonn/Kobori Rats. *Dig Dis. Sci.* **2009**, *54*, 802–810.
23. Koponen, J. M.; Happonen, A. M.; Mattila, P. H.; Törrönen, A. R. Contents of Anthocyanins and Ellagitannins in Selected Foods Consumed in Finland. *J. Agric. Food Chem.* **2007**, *55* (4), 1612–1619.
24. Blomhoff, R.; Carlsen, M. H.; Andersen, L. F.; Jacobs, D. R. Health Benefits Of Nuts: Potential Role of Antioxidants. *Br. J. Nutr.* **2006**, *96* (S2), S52–S60.
25. Cruz-Atonio, F. V.; Saucedo-pompa, S.; Martinez-vázquez, G.; Aguilera, A.; Rodríguez, R.; Aguilar, N. Propiedades Quimicas E Industriales Del Ácido Elágico. *Acta Química Mex.* **2010**, *2* (3).
26. Seeram, N.; Lee, R.; Hardy, M.; Heber, D. Rapid Large Scale Purification of Ellagitannins from Pomegranate Husk, a By-product of the Commercial Juice Industry. *Sep. Purif. Technol.* **2005**, *41* (1), 49–55.
27. Vattem, D. A.; Shetty, K.; Ellagic Acid Production and Phenolic Antioxidant Activity in Cranberry Pomace (*Vaccinium macrocarpon*) Mediated by Lentinus edodes Using a Solid-state System. *Process Biochem.* **2003**, *39* (3), 367–379.
28. de Carvalho, J. C.; Pandey, A.; Oishi, B. O.; Brand, D.; Rodriguez-Léon, J. A.; Soccol, C. R. Relation Between Growth, Respirometric Analysis and Biopigments Production from Monascus by Solid-state Fermentation," *Biochem. Eng. J.* **2006**, *29* (3), 262–269.
29. Pandey, A. Solid-state Fermentation. *Biochem. Eng. J.* **2006**, *13* (2–3), 81–84.
30. Aguilar, C. N.; Contreras-Esquivel, J. C.; Rodríguez-Herrera, R.; Prado-Barragán, L. A.; O. Loera, L. A.; Diferences in Fungal Enzyme Productivity in Submerged and Solid State Cultures. *Food Sci. Biotechnol.* **2004**, *13* (1), 109–113.
31. Aguilar, C. N.; Gutiérrez-Sánchez, G.; Prado-Barragán, L. A.; Rodríguez-Herrera, R.; Martínez-Hernández, J. L.; Contreras-Esquivel, J. C. Perspectives of Solid State Fermentation for Production of Food Enzymes. *Am. J. Biochem. Biotechnol.* **2008**, *1* (4), 354–366.

32. Longo, M. A.; Deive, F. J.; Domínguez, A.; Sanromán, M. Solid-state Fermentation for Food and Feed Application. In *Current Developments in Solid-state Fermentation*; Springer: New York, NY, 2008; pp 379–411.
33. Bellon-Maurel, V.; Orliac, O.; Christen, P. Sensors and Measurements in Solid State Fermentation: A Review. *Process Biochem.* **2003,** *38* (6), 881–896.
34. Domínguez, A.; Costas, M.; Longo, M. A.; Sanromán, A. A Novel Application of Solid State Culture: Production of Lipases by Yarrowia lipolytica. *Biotechnol. Lett.* **2003,** *25* (15), 1225–1229.
35. Aguilera-Carbo, A.; Augur, C.; Prado-Barragan, L. A.; Favela-Torres, E.; Aguilar, C. N. Microbial Production of Ellagic Acid and Biodegradation of Ellagitannins. *Appl. Microbiol. Biotechnol.* **2008,** *78* (2), 189–199.
36. Mata-Gómez, M.; Mussatto, S. I.; Rodríguez, R.; Teixeira, J. A.; Martinez, J. L.; Hernandez, A.; Aguilar, C. N. Gallic Acid Production with Mouldy Polyurethane Particles Obtained from Solid State Culture of *Aspergillus niger* GH1. *Appl. Biochem. Biotechnol.* **2015,** *176* (4), 1131–1140.
37. Wu, M.; Wang, Q.; McKinstry, W. J.; Ren, B. Characterization of a Tannin Acyl Hydrolase from *Streptomyces sviceus* with Substrate Preference for Digalloyl Ester Bonds. *Appl. Microbiol. Biotechnol.* **2014,** *99* (6), 2663–2672.
38. Buenrostro-Figueroa, J.; Ascacio-Valdés, A.; Sepúlveda, L.; De La Cruz, R.l; Prado-Barragán, A.; Aguilar-González, M. A.; Rodríguez, R.; Aguilar, C. N. Potential Use of Different Agroindustrial By-products Assupports for Fungal Ellagitannase Production Undersolid-state Fermentation. *Food Bioprod. Process.* **2013,** *92* (4), 376–382.
39. Ascacio-Valdés, J. A. Estudio de la hidrólisis microbiana de los elagitaninos. 2012.
40. De La Cruz, R.; Aguilera-carbó, A.; Prado-barragán, A.; Rodríguez-herrera, R.; Contreras-esquivel, J.; Aguilar, C.; De Investigación, D.; De, U. A.; De Biotecnología, D.; Autónoma, U.; Antonio, A. Biodegradación Microbiana de Elagitaninos. *BioTecnología* **2011,** *15* (3), 11–18.
41. Huetz, P.; Mavaddat, N.; Mavri, J. Reaction Between Ellagic Acid and an Ultimate Carcinogen. *J. Chem. Inf. Model.* **2005,** *45* (6), 1564–1570.
42. Losso, J. N.; Bansode, R. R.; Trappey, A.; Bawadi, H. A.; Truax, R. In Vitro Antiproliferative Activities of Ellagic Acid. *J. Nutr. Biochem.* **2004,** *15* (11), 672–678.
43. Kelloff, G. J.; Boone, C. W.; Crowell, J. A.; Steele, V. E.; Lubet, R.; Sigman, C. C. Chemopreventive Drug Development: Perspectives and Progress. *Cancer Epidemiol. Biomarkers Prev.* **1994,** *3* (1), 85–98.
44. Wang, N.; Wang, Z. Y.; Mo, S. L.; Loo, T. Y.; Wang, D. M.; Bin Luo, H.; Yang, D. P.; Chen, Y. L.; Shen, J. G.; Chen, J. P. Ellagic Acid, a Phenolic Compound, Exerts Anti-angiogenesis Effects via VEGFR-2 Signaling Pathway in Breast Cancer. *Breast Cancer Res. Treat.* **2012,** *134* (3), 943–955.
45. Venereo Gutiérrez, J. R. Daño oxidativo, radicales libres y antioxidantes. *Rev. Cuba. Med. Mil.* **2002,** *31* (2), 126–133.
46. Zamora S, J. D. Antioxidantes: Micronutrientes En Lucha Por La Salud. *Rev. Chil. Nutr.* **2007,** *34*(1), 17–26.
47. Kilic, I.; Yesiloglu, Y.; Bayrak, Y. Spectroscopic Studies on the Antioxidant Activity of Ellagic Acid. *Spectrochim. Acta - Part A Mol. Biomol. Spectrosc.* **2014,** *130,* 447–452.

132   Green Chemistry and Biodiversity: Principles, Techniques, and Correlations

48. Gil, M. I.; Tomas-Barberan, F. A.; Hess-Pierce, B.; Holcroft, D. M. Kader, A. A. Antioxidant Activity of Pomegranate Juice and Its Relationship with Phenolic Composition and Processing. *J. Agric. Food Chem.* **2000,** *48* (10), 4581–4589.

49. Rizk, H. A.; Masoud, M. A.; Maher, O. W. Prophylactic Effects of Ellagic acid and Rosmarinic Acid on Doxorubicin-Induced Neurotoxicity in Rats. *J. Biochem. Mol. Toxicol. 31* (12), 1–8.

50. Nohynek, L. J.; Alakomi, H., Kähkönen, M. P.; Heinonen, M.; Ilkka, M.; Puupponenpimiä, R. H.; Helander, I. M. Fruit and Vegetables in Cancer Prevention. *Nutr. Cancer* **2006,** *54* (1), 111–142.

51. Machado, T. D. B., Leal, I. C. R.; Amaral, A. C. F.; Santos, K. R. N.; Silva, M. G. Kuster, R. M. Antimicrobial Ellagitannin of *Punica granatum* Fruits. *J. Braz. Chem. Soc.* **2002,** *13* (5), 606–610.

52. Reddy, M. K.; Gupta, S. K.; Jacob, M. R.; Khan, S. I.; Ferreira, D. Antioxidant, Antimalarial, and Antimicrobial Activities of Tannin-rich Fractions, Ellagitannins and Phenolic Acids from *Punica granatum* L. *Planta Med.* **2007,** *73* (5), 461–467.

53. Akiyama, H.; Fujii, K.; Yamasaki, O.; Oono, T.; Iwatsuki, K. Antibacterial Action of Several Tannins Against *Staphylococcus aureus*. *J. Antimicrob. Chemoth.* **2001,** *48* (4), 487–491.

54. Notka, F.; Meier, G.; Wagner, R. Concerted Inhibitory Activities of *Phyllanthus amarus* on HIV Replication In Vitro and Ex Vivo. *Antiviral Res.* **2004,** *64* (2), 93–102.

55. Wong Paz, J. E., Muñiz Márquez, D. B.; Martínez Ávila, G. C. G.; Belmares Cerda, R. E.; Aguilar, C. N. Ultrasound-assisted Extraction of Polyphenols from Native Plants in the Mexican Desert. *Ultrason. Sonochem.* **2015,** *22*, 474–481.

56. Ruibal Brunet, I. J.; Dubed Echevarría, M.; Martínez Luzardo, F.; Noa Romero, E.; Vargas Guerra, L. M.; Santana Romero, J. L. Inhibición de la replicación del virus de inmunodeficiencia humana por extractos de taninos de Pinus caribaea Morelet. *Rev. Cuba. Farm.* **2003,** *37* (2).

57. Serrano, J., Puupponen-Pimiä, R.; Dauer, A. Aura, A. M.; Saura-Calixto, F. Tannins: Current Knowledge of Food Sources, Intake, Bioavailability and Biological Effects. *Mol. Nutr. Food Res.* **2009,** *53*, 310–329.

58. Cheng, H. Y.; Lin, C. C.; Lin, T. C. Antiviral Properties of Prodelphinidin B-2 3′-O-Gallate from Green Tea Leaf. *Antivir. Chem. Chemother.* **2002,** *13* (4), 223–229.

59. Saucedo-Pompa, S.; Rojas-Molina, R.; Aguilera-Carb, A. F.; Saenz-Galindo, A.; de La Garza, H.; Jasso-Cant, D.; Aguilar, C. N. Edible Film Based on Candelilla Wax to Improve the Shelf Life and Quality of Avocado. *Food Res. Int.* **2009,** *42* (4), 511–515.

60. Alvarado, C. J. C.; Galindo, A. S.; Bermdez, L. B.; Berumen, C. P.; vila Orta, C.; Garza, J. A. V.; Cera de Candelilla y sus aplicaciones. *Av. en Quim.* **2012,** *8* (2), 105–110.

61. Przewloka S. R.; Shearer, B. J. The Further Chemistry of Ellagic Acid II. Ellagic Acid and Water-Soluble Ellagates as Metal Precipitants. *Holzforschung* **2002,** *56* (1), 13–19.

62. Aguilar C. N.; Gutiérrez-Sánchez, G. Review: Sources, Properties, Applications and Potential uses of Tannin Acyl Hydrolase. *Food Sci. Technol. Int.* **2001,** *7* (5), 373–382.

63. Doussot, F.; Pardon, P.; Dedier, J.; De Jeso, B. Individual, Species, and Geographic Origin Influence on Cooperage Oak Extractible Content (*Quercus robur* L. and *Quercus petraea* Liebl.). *Analusis* **2000,** *28* (10), 960–965.

Properties and Applications of Ellagic Acid

64. Aqil, F.; Munagala, R.; Jeyabalan, J.; Vadhanam, M. V.; Bioavailability of Phytochemicals and Its Enhancement by Drug Delivery Systems. *Cancer Lett.* **2013**, *334* (1), 133–141.

65. González-Sarrías, A.; García-Villalba, R.; Núñez-Sánchez, M. Á.; Tomé-Carneiro, J.; Zafrilla, P.; Mulero, J.; Tomás-Barberán, F. A.; Espín, J. C.; Identifying the Limits for Ellagic Acid Bioavailability: A Crossover Pharmacokinetic Study in Healthy Volunteers after Consumption of Pomegranate Extracts. *J. Funct. Foods* **2015**, *19*, 225–235.

66. Ting, Y.; Jiang, Y.; Ho, C. T.; Huang, Q. Common Delivery Systems for Enhancing in Vivo Bioavailability and Biological Efficacy of Nutraceuticals. *J. Funct. Foods* **2014**, *7* (1), 112–128.

67. Seeram, N. P.; Lee, R.; Heber, D.; Bioavailability of Ellagic Acid in Human Plasma after Consumption of Ellagitannins from Pomegranate (*Punica granatum* L.) juice. *Clin. Chim. Acta* **2004**, *348* (1–2), 63–68.

68. Mertens-talcott, S. U.; Jilma-stohlawetz, P.; Rios, J.; Hingorani, L.; Derendorf, H. Absorption , Metabolism , and Antioxidant Effects of Pomegranate (*Punica granatum* L .) Polyphenols after Ingestion of a Standardized Extract in Healthy Human Volunteers *J. Agric. Food Chem.* **2006**, *54*, 8956–8961.

69. Romo-Vaquero, M.; García-Villalba, R.; González-Sarrías, A.; Beltrán, D.; Tomás-Barberán, F. A.; Espín, J. C.; Selma, M. V., Interindividual Variability in the Human Metabolism of Ellagic Acid: Contribution of Gordonibacter to Urolithin Production. *J. Funct. Foods* **2015**, *17*, 785–791.

70. Ascacio-Valdés, J.; Buenrostro, J. J.; De la Cruz, R.; Sepúlveda, L.; Aguilera, A. F.; Prado, A.; Contreras, J. C.; Rodríguez, R.; Aguilar, C. N. Fungal Biodegradation of Pomegranate Ellagitannins. *J. Basic Microbiol.* **2014**, *54* (1), 28–34.

71. Aguilera-Carbo, A.; Hernández, J. S.; Augur, C.; Prado-Barragan, L. A.; Favela-Torres, E.; Aguilar, C. N. Ellagic Acid Production from Biodegradation of Creosote Bush Ellagitannins by *Aspergillus niger* in Solid State Culture. *Food Bioproc. Technol.* **2009**, *2* (2), 208–212.

72. Sepúlveda, L.; Aguilera-carbó, A.; Ascacio-valdés, J. A.; Rodríguez-herrera, R.; Martínez-hernández, J. L.; Aguilar, C. N. Optimization of Ellagic Acid Accumulation by *Aspergillus niger* GH1 in Solid State Culture Using Pomegranate Shell Powder as a Support. *Process Biochem.* **2012**, *47* (12), 2199–2203.

73. De la Cruz, R. Optimización de las condiciones de producción de una elagitanasa fúngica, 2012.

Green Chemistry and Biodiversity Principles, Techniques, and Correlations   A

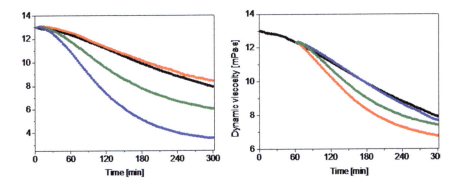

FIGURE 10.1 Time dependent changes in dynamic viscosity of hyaluronan solution exposed to oxidative degradation by Cu(II) ions and ascorbic acid (black curve) and in the presence of *Punica granatum* at volume: 50 (red curve), 200 (green curve), and 500 µL (blue curve) added before hyaluronan degradation begins (left panel) or 1 h later (right panel).

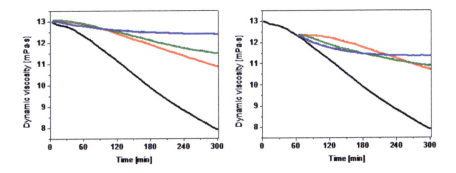

FIGURE 10.2 Time dependent changes in dynamic viscosity of hyaluronan solution exposed to oxidative degradation by Cu(II) ions and ascorbic acid (black curve) and in the presence of *Peganum harmala* at volume: 200 (red curve), 500 (green curve), and 1000 µL (blue curve) added before hyaluronan degradation begins (left panel) or 1 h later (right panel).

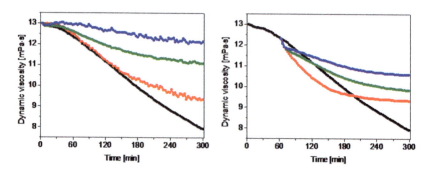

FIGURE 10.3 Time-dependent changes in dynamic viscosity of hyaluronan solution exposed to oxidative degradation by Cu(II) ions and ascorbic acid (black curve) and in the presence of *Dianthus caryophyllus* at volume: 200 (red curve), 500 (green curve), and 1000 μL (blue curve) added before hyaluronan degradation begins (left panel) or 1 h later (right panel).

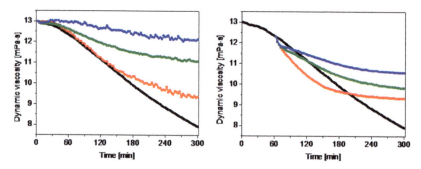

FIGURE 10.4 Time dependent changes in dynamic viscosity of hyaluronan solution exposed to oxidative degradation by Cu(II) ions and ascorbic acid (black curve) and in the presence of *Vitis vinifera* at volume: 50 (red curve), 100 (blue curve), and 500 μL (green curve) added before hyaluronan degradation begins (left panel) or 1 h later (right panel; Table 10.1).

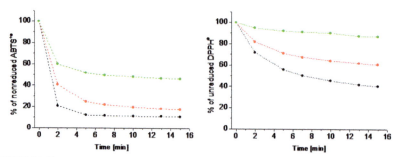

FIGURE 10.5 Percentage of reduction of ABTS$^{•+}$ (right panel) and DPPH$^{•}$ (left panel) by using the *Punica granatum* extract at concentrations 1 (green curve), 3 (red curve), and 5 mg/mL (black curve).

Green Chemistry and Biodiversity Principles, Techniques, and Correlations   C

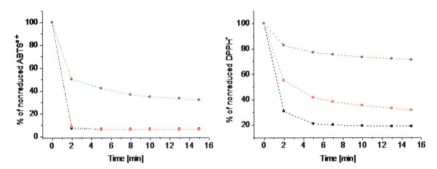

FIGURE 10.6 Percentage of reduction of ABTS$^{•+}$ (right panel) and DPPH$^{•}$ (left panel) by using the *Dianthus caryophyllus* extract at concentrations 1 (green), 3 (red), and 5 mg/mL (black).

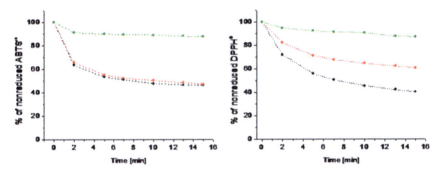

FIGURE 10.7 Percentage of reduction of ABTS$^{•+}$ (right panel) and DPPH$^{•}$ (left panel) by using the extract of *Vitis vinifera* at concentrations 1 (green), 3 (red), and 5 mg/mL (black).

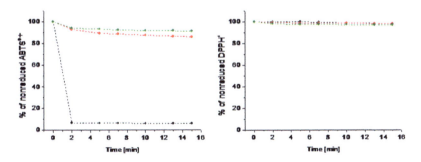

FIGURE 10.8 Percentage of reduction of ABTS$^{•+}$ (right panel) and DPPH$^{•}$ (left panel) by using the extract of *Peganum harmala* at concentrations 1 (green), 3 (red), and 5 mg/mL (black).

D   Green Chemistry and Biodiversity: Principles, Techniques, and Correlations

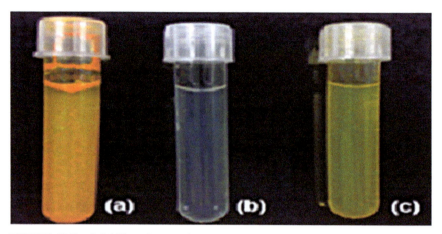

FIGURE 12.2 Solubility of curcumin (a) curcumin in water, (b) O-CMC NPs, and (c) curcumin-OCMC NPs. © 2016 Korean Institute of Chemical Engineers, Seoul, Korea.

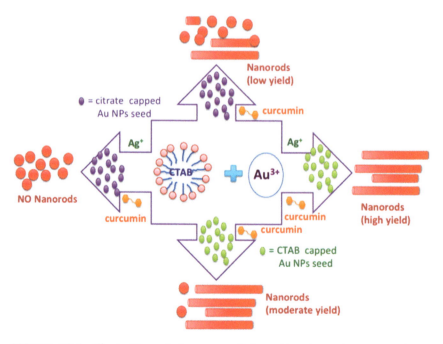

FIGURE 12.5 Illustration of Au nanorod formation by using curcumin as secondary reducing agent through seed mediated method.
Source: **Reprinted** with permission from Ref. [43]. © 2015 American Chemical Society.

CHAPTER 10

Antioxidative Properties of *Punica granatum, Peganum harmala, Dianthus caryophyllus*, and *Vitis vinifera* Extracts Against Free Radicals

KATARÍNA VALACHOVÁ[1*], ELSAYED E. HAFEZ[2], MILAN NAGY[3], and LADISLAV ŠOLTÉS[1]

[1]*Centre of Experimental Medicine, Institute of Experimental Pharmacology and Toxicology, Slovak Academy of Sciences, Bratislava, Slovakia*

[2]*Plant Protection and Biomolecular Diagnosis Department, Arid Lands Cultivation Research Institute, City of Scientific Research and Technological Applications, New Borg El-Arab City, Alexandria 21934, Egypt*

[3]*Department of Pharmacognosy and Botany, Faculty of Pharmacy, Comenius University, Bratislava, Slovakia*

Corresponding author. E-mail: kataria.valachova@savba.sk

ABSTRACT

Extracts of *Punica granatum, Peganum harmala, Dianthus caryophyllus,* and *Vitis vinifera* are known for their protective-antioxidative effects in various disorders. We examined the extracts ability to scavenge reactive oxygen species such as hydroxyl radicals, alkoxy-, and peroxy-type radicals by means of rotational viscometry. Determination of IC_{50} values and the capability of the extracts to reduce $ABTS^{\bullet+}$ and $DPPH^{\bullet}$ radicals were assessed by the ABTS and DPPH methods. Spectrophotometrically, we determined the content of tannins in the extracts. Results from rotational

136 Green Chemistry and Biodiversity: Principles, Techniques, and Correlations

viscometry show that all examined extracts dose-dependently attenuated free-radical degradation of high-molar-mass hyaluronan. *Punica granatum* extract was the most effective in the ABTS and DPPH assays.

10.1 INTRODUCTION

Phytochemicals, especially phenolics are suggested to be the major bioactive compounds with health benefits. Clinical trials and epidemiological studies have established an inverse correlation between the intake of dietary antioxidants and the occurrence of oxidative stress related diseases. The bioactivity of phenolics may be due to their ability to inhibit lipoxygenase, to chelate metals, to scavenge free radicals, or to prevent adverse effects of reactive oxygen and/or nitrogen species on normal physiological function in humans (Satheesh et al., 2013).

Pomegranate (*Punica granatum* L.) fruit is well-known for its nutritional and sensory properties. It grows in most tropical and subtropical countries and belongs to the Lythraceae family. Edible parts of this fruit are fresh seeds containing around 80% of juice and 20% of seed. The fruit is rich in minerals, pigments, galloylglucose, ellagic acid, alkaloids, glycosides, resins, volatile oils, gums, and tannins especially anthocyanins (glucosides of delphinidin and cyanidin), which have chemopreventive, antimutagenic, antibacterial, antihypertensive, antiatherogenic, and antioxidative properties. The fruit is also effective in reduction in liver injury (Bhandary et al., 2012; Rummun et al., 2013; Barman et al., 2014; Zhao et al., 2016). Pomegranate can be used in treatment and prevention of cardiovascular diseases, diabetes, dental conditions, and protection from ultraviolet radiation. Moreover, it is beneficial in infant brain ischemia, Alzheimer's disease, male infertility, arthritis, and obesity (Kumar et al., 2012; Zhao et al., 2016).

Consumption of pomegranate juice has been associated with a decrease in a level of inflammatory biomarkers and oxidation of proteins and lipids in a randomized placebo-controlled trial. The beneficial effect of pomegranate juice was reported in an initial phase II clinical trial in patients suffering from prostate cancer. The health benefits of pomegranate can be attributed to the pluripharmacological effects of the secondary metabolites more specifically its polyphenolic compounds present in relatively high concentrations. The phytophenolic compositions divide in the edible and nonedible parts of the plants and have been widely investigated (Rummun et al., 2013).

Antioxidative Properties   137

Peganum harmala L. (Syrian rue) is a wild-growing flowering plant belonging to the family Nitrariaceae and is abundant in the Middle East and North Africa. It is claimed to be an important medicinal plant. *Peganum harmala* seeds are known to possess hypothermic, antimicrobial, hepatoprotective as well as antioxidative, hypoglycemic, cytoprotective, antihelmintic, antispasmodic, antipyretic, antiprotozoal, non-nephrotoxic, antivirous, and antihypersensitive effects. It is also effective in the treatment of dermatosis (Tse et al., 1991; Berrougui et al., 2006; Hamden et al., 2008; Soliman et al., 2011; Ahmed et al., 2013; Moloudizargari et al., 2013; Karam et al., 2016; Komeili et al., 2016). *Peganum harmala* is rich in β-carboline alkaloids (harmine, harmaline, harmalol, harman) and quinazoline ones (vasicine and vasicinon). These alkaloids have a broad spectrum of potent therapeutic activities such as anticancer, analgesic, and antinociceptive. They inhibit monoamine oxidase and platelet aggregation. They bind to benzodiazepine receptors and have immunomodulatory, vasorelaxant, anxiolytic, cytotoxic, anti-bacterial, antiplasmodial, antileishmanial, immunomodulatory, temperature-lowering activities, and cardiovascular actions. Recent research reported that the alkaloids showed potential inhibitory effects on the acetylcholinesterase and behavioral effects (Tse et al., 1991; Berrougui et al., 2006; Hamden et al., 2008; Liu et al., 2013; Karam et al., 2016; Komeili et al., 2016; Moradi et al., 2017). The above mentioned alkaloids are known for their multi-enzymes inhibitions including cyclin-dependent kinases, *N*-acetyltransferase, and Na–K ATPase.

Carbolines structurally related to harmala alkaloids have also been found endogenously in mammalian tissues, including the central nervous system, liver, platelets, plasma, and urine. There were shown cardiovascular actions of harmala alkaloids such as harmin, harmalol, and harmaline, which reduced systemic arterial blood pressure and total vascular resistance (Tse et al., 1991). Moreover, *Peganum harmala* was shown to be concerned on cardiovascular actions, and DNA topoisomerase inhibition in cancerous cell-lines (Hamden et al., 2008). Previous pharmacological studies on the compounds from this plant revealed that β-carboline and quinazoline alkaloids were the main physiological active ingredients as well as the causations of toxicity and side effects. For example, essential tremor was a drawback to their usage in the clinical therapy. Interestingly, harmine was once used as an anti-parkinsonian agent however it was soon abandoned for its disappointing effectiveness.

By phytochemical, analysis of carnation (*Dianthus caryophyllus* L., family Caryophyllaceae) was shown to contain alkaloids, cyanidin, coumarins, cyanogenic glycosides, isosalipurporoside, pelagonidin, saponins, and volatile oil. Pharmacological studies revealed that the plant has anticancer, antiviral, antibacterial, antifungal, antispasmodic, insecticidal, repellent, antiseptic, antioxidative, anesthetic, and analgesic effects. It is traditionally used in China, Japan, and South Korea in perfumery, where 500 kg of flowers produce 100 g of volatile oil. Further, it is used for the treatment of a sore throat, gum infections, gastro-intestinal, and sight disorders. Moreover, it is cardiotonic and diaphoretic. It was prescribed in Europe as a herbal medicine to treat coronary and nervous disorders. For a long time, the carnation was used as medicine and spice. Essential oil of carnation was applied to improve memory and to restore forces. It was used to heal wounds, relieve dizziness, and support appetite (Al-Snafi, 2017).

It is used as tea, thereby it may help reduce stress and fatigue. Carnation oil has therapeutic benefits for the treatment of skin rashes, rosacea, and eczema. Moreover, it can be used as a conditioner for skin and to minimize facial wrinkles. Carnation contains components that reduce inflammation, swelling, and can help restore natural hormonal balances in women. Carnation has long been used to reduce muscle tension in uterine tissues and to treat endometriosis (http://www.livestrong.com/article/70178-medicinal-uses-carnations/2016).

The flower petals can be candied and used as a garnish in salads, for flavoring fruit and fruit salads. The plant is quite rich in saponins. The leaves of carnation can be used for production of soaps (http://www.naturalmedicinalherbs.net/herbs/d/dianthus-caryophyllus=carnation.php, 2018).

Grape (*Vitis vinifera* L., family Vitaceae) belongs to the world's largest fruit crops. Grape seed is a complex matrix containing approximately 40% of fiber, 16% of oil, 11% of proteins, and 7% of complex phenolics in addition to saccharides, salts, and minerals. Phenolic compounds include proanthocyanidins, which are known for their therapeutic potentials and pharmacological activities. Other polyphenols present in grape seed are (+)-catechin, (-)-epicatechin, (-)-epicatechin-3-*O*-gallate, dimeric, trimeric, and tetrameric procyanidins, which possess protective properties against reactive oxygen species and oxidative stress as well as have anti-inflammatory, antibacterial, antiviral, antimutagenic, antioxidative, and antidiabetic activities (Kim et al., 2006; Suwannaphet et al., 2010; Choi et al., 2012).

Antioxidative Properties 139

Grape seeds also contain α-, β-, and γ-tocopherols as well as α-and β-tocotrienols, which possess strong antioxidative activity. Besides free radicals scavenging, they are also capable to inhibit lipid peroxidation (Kim et al., 2006; Choi et al., 2012). Grape seed extract prevents heart diseases and reduces infarct size in experimental cardiac ischemia. It also protects liver from acetaminophen-induced damage and puromycin-induced nephrosis (Dulundu et al., 2007; Feringa et al., 2011).

Commercial preparations of grape seed extract are marketed worldwide as food supplements. The cancer chemoprevention and anticancer potential of grape seed extract has been well reviewed including skin, colorectal, prostate, breast, lung, and gastric cancers. A low concentration (2.5 µg/mL) the grape seed extract was reported to inhibit the micronuclei frequency and generation of reactive oxygen species in a lymphocyte culture, demonstrating that its antioxidant property has a protective effect during oxidative stress. On the other hand, its high concentrations (25–100 µg/mL) showed cytotoxicity or antiproliferation of human bladder, colorectal, and breast cancer cell lines (Yen et al., 2015). Grape seed extract is commercially available and is prepared from the seed of grapes. Oral grape seed extract is typically administered in 50- or 100-mg capsules or tablets. Grape seed extract is also found in wine, whereas red wines contain substantially more grapes than white wines (177 mg/L compared to 8.75 mg/L). Grape seed extract protects elastin, collagen, and hyaluronic acid within the skin by blocking enzymes that may degrade them and/or disrupt their chemical structure. In this way, grape seed extract supports the skin young appearance. It prevents skin damage derived from sunlight, pollution, X-rays, cigarette smoke, and even stress (Yuan et al., 2012).

10.2 MATERIAL AND METHODS

10.2.1 CHEMICALS

Hyaluronan (sodium salt) coded HA15M-5 of the average molar mass 1.93 MDa was purchased from Lifecore Biomedical Inc., Chaska, MN, USA. Analytical purity grade NaCl and $CuCl_2 \cdot 2H_2O$ were purchased from Slavus Ltd., Bratislava, Slovakia. *Peganum harmala* and *Dianthus caryophyllus* (powders from leaves), *Punica granatum, Vitis vinifera* (powders from seeds) extracts were obtained from City of Scientific Research and Technological Applications, Alexandia, Egypt. L-Ascorbic acid and potassium persulfate ($K_2S_2O_8$ p.a. purity, max. 0.001% nitrogen) were the products of Merck KGaA, Darmstadt, Germany. 2,2′-Azinobis-(3-ethylbenzothiazoline-6-sulfonic acid) diammonium salt (ABTS; purum, >99%) was from Fluka, Germany. 2,2-Diphenyl-1-picrylhydrazyl (DPPH; 95%), hide powder, sodium carbonate, Folin & Ciocalteu's phenol reagent were from Sigma-Aldrich, Germany. Methanol was the product of MikroChem, Pezinok, Slovakia. Redistilled deionized high quality grade water, with conductivity of <0.055 µS/cm, was produced using the TKA water purification system from Water Purification Systems GmbH, Niederelbert, Germany.

10.3 EXPERIMENTAL

10.3.1 PREPARATION OF STOCK AND WORKING SOLUTIONS

Hyaluronan HA15M-5 (14 mg) was dissolved in 0.15 mmol/L aqueous NaCl solution for 24 h in the dark. Hyaluronan sample solutions were prepared in two steps: first, 4.0 mL and after 6 h 3.9 mL of 0.15 mmol/L NaCl was added in the absence of the plant extract. NaCl (0.15 mmol/L, 3.85–2.9 mL) was added in the presence of a plant extract. Solutions of ascorbate (16 mmol/L), and cupric chloride solution (160 µmol/L) were prepared also in 0.15 mol/L aqueous NaCl.

All aqueous extracts were prepared as follows: The plant dry powdered extracts (50 mg) were leached for 15 min in boiled distilled water (40 mL). Then the solutions were cooled down, filtered, and completed with distilled water to the volume of 50 mL. Concentrations of the extracts were 3.8 mg/mL (*Vitis vinifera*), 4.4 mg/mL (*Dianthus caryophyllus*), 3.4 mg/mL (*Punica granatum*) and 2.9 mg/mL (*Peganum harmala*), respectively.

10.3.2 DETERMINATION OF TANNINS CONTENT

The aqueous solutions in the volume of 25 mL (0.2 mg/mL) of *Peganum harmala, Punica granatum, Dianthus caryophyllus*, and *Vitis vinifera*, respectively, were prepared.

2 mL of each aqueous extract was mixed with 1 mL of Folin & Ciocalteu's phenol reagent and 10 mL of distilled water. The reaction mixture was made up to the total volume 25 mL with 29% sodium carbonate. Absorbance (A_1) at 760 nm of the samples was recorded after 30 min.

Antioxidative Properties 141

The content of tannins (A_2) was determined as follows: 10 mL of each aqueous extract was mixed with 0.1 g of hide powder and left on a shaker for 60 min. Then the solution was filtered and 5 mL of the filtered solution was mixed with 1 mL of Folin and Ciocalteu's phenol reagent and 10 mL of distilled water. Reaction mixture was made up to total volume 25 mL with 29% sodium carbonate. Absorbance (A_2) at 760 nm of the samples was recorded after 30 min.

Percentage of the content of tannins in the extracts was calculated according to the following equation:

$$\% = 62.5 * (A_1 - A_2) * m_2/A_3 * m_1,$$

where A_1—absorbance of the measured solution when determining the total polyphenol content,

A_2—absorbance of the measured solution when determining tannins, that is, the polyphenol content not absorbed by hide powder,

A_3—absorbance of pyrogallol as a reference ($A_3 = 0.35$),

m_1—weight of the sample (g), and

m_2—weight of pyrogallol (g).

10.3.3 *ROTATIONAL VISCOMETRY (UNINHIBITED/INHIBITED HYALURONAN DEGRADATION)*

The procedure was performed as described in Banasova et al. (2012), Valachova et al. (2009, 2010, 2015), Hrabarova et al. (2009), and Topolska et al. (2014, 2015). Degradation of high-molar-mass hyaluronan was induced in vitro by Weissberger's biogenic oxidative system comprising 100 µmol/L ascorbate and 1 µmol/L $CuCl_2$, applied under aerobic conditions. The reaction mixture was transferred into the Teflon® cup reservoir of a Brookfield LVDV-II+PRO digital rotational viscometer (Brookfield Engineering Labs., Inc., Middleboro, MA, USA) and changes in dynamic viscosity of the hyaluronan in vitro solution were recorded at 25.0°C ± 0.1°C in 3 min intervals for 5 h. Two experimental regimes were applied for assessing the influence of the plant extracts on hyaluronan degradation. Firstly, each drug was added to the reaction mixture 30 s before the addition of ascorbic acid, which initiates oxidative degradation of hyaluronan by producing •OH radicals. And secondly, each extract was added to the reaction mixture 1 h later, when production of alkoxy-/peroxy-type radicals predominates.

10.3.4 ABTS ASSAY—KINETICS AND DETERMINATION OF IC$_{50}$ VALUES

The first step of the standard ABTS assay was preparation of the aqueous solution of ABTS$^{\bullet+}$ cation radical. ABTS$^{\bullet+}$ was prepared 24 h before the measurements at room temperature as follows: ABTS aqueous stock solution (7 mmol/L) was mixed with $K_2S_2O_8$ aqueous solution (2.45 mmol/L) in the equivolume ratio. On the next day, 1 mL of the resulting solution was diluted with distilled water to the final volume of 60 mL (Re et al., 1999). The aqueous reagent in the volume of 250 μL was added to 2.5 μL of the aqueous solutions of the plant extracts. Absorbance at 734 nm of samples was recorded after 6 min.

Determination of IC$_{50}$ values: The aqueous solutions of all plant extracts were prepared at concentration 5 mg/mL.

Kinetic measurements: The stock solutions of the plant extracts used for monitoring the reduction of ABTS$^{\bullet+}$ cation radical were at concentrations 1, 3, and 5 mg/mL.

10.3.5 DPPH ASSAY—DETERMINATION OF IC$_{50}$ VALUES

The first step of a standard DPPH assay was preparation of DPPH$^{\bullet}$ radical as follows: 2,2-diphenyl-1-picrylhydrazyl (1.1 mg) was dissolved in 50 mL of distilled methanol. The DPPH$^{\bullet}$ solution in the volume of 225 μL was added to 25 μL of the methanol solution of the plant extracts. Absorbance at 517 nm of samples was recorded after 30 min.

The solutions of all plant extracts were prepared as follows: The plant extracts (8 mg/mL) were dissolved in methanol (25 mL). The solutions were sonicated for 15 minutes and dried in a desiccator to reach a final concentration of 2.4, 2.1, 2.0, and 1.5 mg/mL for the extracts of *Punica granatum*, *Dianthus caryophyllus*, *Peganum harmala*, and *Vitis vinifera*, respectively.

In both assays the measurements were performed in the quadruplicate in 96-well Greiner UV-Star microplates (Greiner-BioOne GmbH, Germany) by using the Tecan Infinite M 200 reader (Tecan AG, Austria).

Kinetic measurements: The aqueous solutions of the plant extracts used for monitoring the reduction of DPPH$^{\bullet}$ radical were at concentrations 1, 3, and 5 mg/mL.

10.4 RESULTS AND DISCUSSION

As seen in Figure 10.1, hyaluronan was firstly subjected to degradation by WBOS, which corresponds to decrease in dynamic viscosity of the hyaluronan solution by 5.0 mPa.s within 5 h (black curve, the reference). The extract of *Punica granatum* added in the volume 50 µL had no protective effect against •OH radical-induced hyaluronan degradation and the red curve corresponds to the reference curve. Moreover, the addition of the extract in the volumes 200 and 500 µL facilitated hyaluronan degradation and compared to the reference, the decrease in dynamic viscosity of the hyaluronan solutions were 6.9 and 9.4 mPa.s, respectively. The reason might be the presence of vitamin C in this extract, which in our system acts as a pro-oxidant. The presence of vitamin C in the *Punica granatum* extract was described in several papers (Opara et al., 2009; Dumlu et al., 2007; Amararatne et al., 2012; Ismail et al., 2014; Akkiraju et al., 2016).

Right panel shows results with the addition of the extract 1 h later, where alkoxy- and peroxy-type radicals rather than •OH radicals are prevailed. The extract added in the given volumes did not prevent hyaluronan degradation. However, the higher amount of the extract was added, the lower decrease in dynamic viscosity of the hyaluronan solution was monitored.

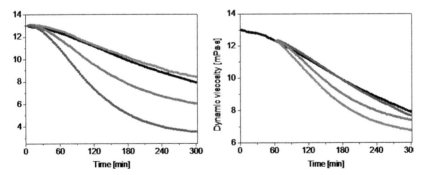

FIGURE 10.1 (See color insert.) Time dependent changes in dynamic viscosity of hyaluronan solution exposed to oxidative degradation by Cu(II) ions and ascorbic acid (black curve) and in the presence of *Punica granatum* at volume: 50 (red curve), 200 (green curve), and 500 µL (blue curve) added before hyaluronan degradation begins (left panel) or 1 h later (right panel).

The extract of *Peganum harmala* (Fig. 10.2, left panel) was shown to dose-dependently inhibit •OH radical-induced degradation of hyaluronan. The reached decreases in dynamic viscosity of the hyaluronan solutions were 2.1, 1.47, and 0.58 mPa.s for the extract at volumes 200, 500, and 1000 µL, respectively. In the second experimental regime, the extract effectively inhibited alkoxy-/peroxy-type radical hyaluronan degradation; however, independently of the added amount of the extract. The decreases in dynamic viscosity of the hyaluronan solution were in a range from 1.61 to 0.96 mPa.s (right panel).

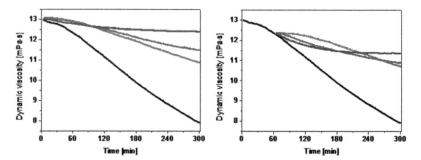

FIGURE 10.2 (See color insert.) Time dependent changes in dynamic viscosity of hyaluronan solution exposed to oxidative degradation by Cu(II) ions and ascorbic acid (black curve) and in the presence of *Peganum harmala* at volume: 200 (red curve), 500 (green curve), and 1000 µL (blue curve) added before hyaluronan degradation begins (left panel) or 1 h later (right panel).

The ability of the *Peganum harmala* extract to scavenge •OH radicals was reported also by Mekki (2014).

Results in Figure 10.3 show a dose-dependent inhibitory activity of *Dianthus caryophyllus* extract against OH radical-mediated hyaluronan degradation. Dynamic viscosity of the hyaluronan solutions was decreased by 3.63, 1.91, and 0.85 mPa.s when adding the extract in volumes 200, 500, and 1000 µL, respectively. The carnation extract was also effective against alkoxy-/peroxy-type radical- induced hyaluronan degradation. The dynamic viscosity of the hyaluronan solutions was decreased by 2.98, 2.48, and 1.73 mPa.s within 5 h.

As seen in Figure 10.4, the *Vitis vinifera* extract diminished hyaluronan degradation induced by •OH and alkoxy-/peroxy-type radicals in all used concentrations.

Antioxidative Properties

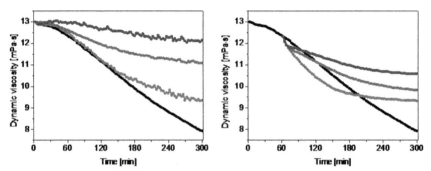

FIGURE 10.3 **(See color insert.)** Time-dependent changes in dynamic viscosity of hyaluronan solution exposed to oxidative degradation by Cu(II) ions and ascorbic acid (black curve) and in the presence of *Dianthus caryophyllus* at volume: 200 (red curve), 500 (green curve), and 1000 μL (blue curve) added before hyaluronan degradation begins (left panel) or 1 h later (right panel).

Protective effects of the extract against hydroxyl radicals were demonstrated also in numerous papers (Baghi et al., 1997; Yamaguchi et al., 1999; Sreemantula et al., 2005; Li et al., 2008; Mandic et al., 2008; El-Beltagi et al.; 2016; Pirincciouglu et al., 2017).

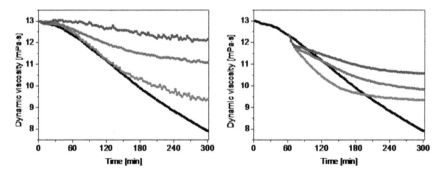

FIGURE 10.4 **(See color insert.)** Time dependent changes in dynamic viscosity of hyaluronan solution exposed to oxidative degradation by Cu(II) ions and ascorbic acid (black curve) and in the presence of *Vitis vinifera* at volume: 50 (red curve), 100 (blue curve), and 500 μL (green curve) added before hyaluronan degradation begins (left panel) or 1 h later (right panel; Table 10.1).

TABLE 10.1 IC$_{50}$ Values of the Plant Extracts by Using the ABTS and DPPH Assay and Content of Tannins.

Extract sample	ABTS IC$_{50}$ (µg/mL)	DPPH IC$_{50}$ (µg/mL)	Tannins content (%)
Peganum harmala	194.3 ± 4.5	Not detected	1.5
Punica granatum	5.2 ± 0.4	2.85 ± 0.4	4.2
Vitis vinifera	47.7 ± 1.5	25.2 ± 0.8	1.2
Dianthus caryophyllus	22.6 ± 0.3	16.1 ± 0.4	2.9

The *Punica granatum* extract was shown to be the most effective sample with the IC$_{50}$ value reaching 5.2 ± 0.4 µg/mL in the ABTS assay. In contrast, the least effective extract was of *Peganum harmala*. In the DPPH assay, the most effective sample was again shown to be *Punica granatum* extract. A bit less effective antioxidants were *Vitis vinifera* and *Dianthus caryophyllus*. No value was determined for the extract of *Peganum harmala*. Of the examined extracts the highest content of tannins was determined in the *Punica granatum* extract, namely 4.2 mass percent.

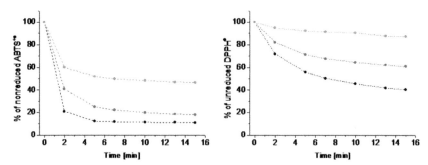

FIGURE 10.5 **(See color insert.)** Percentage of reduction of ABTS$^{•+}$ (right panel) and DPPH$^{•}$ (left panel) by using the *Punica granatum* extract at concentrations 1 (green curve), 3 (red curve), and 5 mg/mL (black curve).

The ability of the *Punica granatum* extract was examined by the ABTS and DPPH assays (Fig. 10.5). This extract showed good radical scavenging properties since the percentages of non-reduced ABTS$^{•+}$ cation radical were 11.0%, 18.0%, and 46.5% when examining the extract at concentrations 1, 3, and 5 mg/mL (left panel). A lower effect of the *Punica granatum* extract was evidenced using the DPPH assay (right panel). The amounts of non-reduced DPPH• radical were 40.2%, 60.7%, 87.2% at concentrations 1, 3, and 5 mg/mL, respectively.

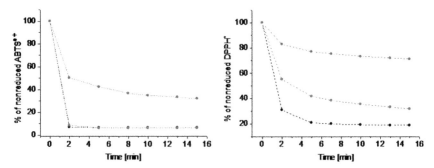

FIGURE 10.6 (See color insert.) Percentage of reduction of ABTS•+ (right panel) and DPPH• (left panel) by using the *Dianthus caryophyllus* extract at concentrations 1 (green), 3 (red), and 5 mg/mL (black).

As shown in Figure 10.6, left panel, the percentages of non-reduced ABTS•+ within 15 min were 6.3%, 6.2%, and 32 % when examining the extract at concentrations 5, 3, and 1 mg/mL, respectively.

On the other hand, the extract possesses lower radical scavenging activity in the DPPH assay. After the 15-min measurement the amounts of the non-reduced DPPH• radical were 18.8%, 31.8%, and 71.4% using the extract at concentrations 5, 3, and 1 mg/mL, respectively.

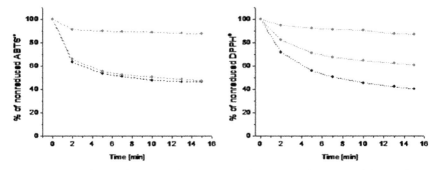

FIGURE 10.7 (See color insert.) Percentage of reduction of ABTS•+ (right panel) and DPPH• (left panel) by using the extract of *Vitis vinifera* at concentrations 1 (green), 3 (red), and 5 mg/mL (black).

As shown in Figure 10.7, left panel, the extract of *Vitis vinifera* at concentrations 1 and 3 mg/mL (red and black curves) the amount of non-reduced ABTS was ca. 50 %. On the other hand, this extract at concentration 5 mg/mL had only a slight effect on reduction of ABTS$^{•+}$ cation radical. In this case, the percentage of non-reduced radical was 87.5%. As seen in right panel, this extract dose-dependently reduced DPPH$^{•}$ radical. Percentages of non-reduced DPPH$^{•}$ radical within 15 min were 40.2%, 60.7%, and 87.2% for the extract used at concentrations 1, 3, and 5 mg/mL, respectively.

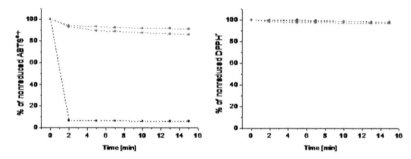

FIGURE 10.8 (See color insert.) Percentage of reduction of ABTS$^{•+}$ (right panel) and DPPH$^{•}$ (left panel) by using the extract of *Peganum harmala* at concentrations 1 (green), 3 (red), and 5 mg/mL (black).

Figure 10.8 shows the results of the effect of *Peganum harmala* extract on decolorization of both ABTS$^{•+}$ and DPPH$^{•}$ radicals solutions. As evident in the left panel, this extract at the highest concentration almost completely reduced to ABTS$^{•+}$ cation radical. The amount of this non-reduced radical was 6.2%. In contrast, the extract examined at lower concentrations was almost ineffective. Concerning to the right panel, this extract at all examined concentrations almost did not reduce DPPH$^{•}$ radical at all. To explain the observed differences on working with the two assays one should take into account that the examined samples are multicomponent mixtures which may differently dissociate in water (the ABTS assay) and methanol (the DPPH assay).

10.5 CONCLUSION

Based on the results, it can be stated that the examined extracts dose-dependently diminished hyaluronan degradation. In the ABTS and DPPH assays, the *Punica granatum* extract was the most effective sample.

KEYWORDS

- biopolymer oxidative degradation
- glycosaminoglycans
- plant extracts
- reactive oxygen species
- rotational viscometry

REFERENCES

Ahmed, H.; Abu El Zahab, H.; Alswiai, G. Purification of Antioxidant Protein Isolated from Peganum Harmala and its Protective Effect Against CCl$_4$ Toxicity in Rats. *Turk. J. Biol.* **2013**, *37*, 39–48.

Akkiraju, P. C.; Suryawanshi, D. D.; Jawakekar, A. J.; Tambe H. S.; Mamillapalli, S. Phytochemical Analysis and HPLC Study of Vitamin-C from *Punica Granatum L.* Aarakta Variety of India. *J. Med. Plants Stud.* **2016**, *4* (6), 09.

Al-Snafi, A. E. Chemical Contents and Medical Importance of *Dianthus Caryophyllus*—A Review. *IOSR J. Pharm.* **2017**, *7* (3), 61–71.

Amararatne, D. I. M.; Weerakkody, W. A. P.; Jayakody, J. A. L. P. Bioactive Properties of Fruit Juice of Pomegranate (*Punica Granatum*) Grown in Dry Regions of Sri Lanka. *Trop. Agric. Res.* **2012**, *23* (4), 370–375.

Arun, N.; Singh, D. P. *Punica Granatum*: A Review on Pharmacological and Therapeutic Properties. *Int. J. Pharm. Sci. Res.* **2012**, *3* (5), 1240–1245.

Baňasová, M.; Sasinková, V.; Mendichi, R.; Perečko, T.; Valachová, K.; Juránek, I.; Šoltés, L. *Neuroendocrinol. Lett.* **2012**, *33* (Supp.3), 151–154.

Bagchi, D.; Garg, A.; Krohn, R. L.; Bagchi, M.; Tran, M. X.; Stohs, S. J. Oxygen Free Radical Scavenging Abilities of Vitamins C and E, and A Grape Seed Proanthocyanidin Extract In Vitro. *Res. Commun. Mol. Pathol. Pharmacol.* **1997**, *95* (2), 179–189.

Barman, K.; Asrey, R.; Pal, R. K.; Kaur, C.; Jha, S. K. Influence of Putrescine and Carnauba Wax on Functional and Sensory Quality of Pomegranate (*Punica Granatum L.*) Fruits During Storage. *J. Food Sci. Technol.* **2014**, *51* (1), 111–117.

Berrougui, H.; Isabelle, M.; Cloutier, M.; Hmamouchi, M.; Khalil, A. Protective Effects of Peganum Harmala L. Extract, Harmine and Harmaline Against Human Low-Density Lipoprotein Oxidation. *J. Pharm. Pharmacol.* **2006**, *58*, 967–974.

Bhandary, K.; Kumari, S. N.; Bhat, V. S.; Sharmila, K. P.; Beka, M. P. Preliminary Phytochemical Screening of Various Extracts of *Punica granatum* Peel, Whole Fruit and Seeds. *Nitte Univers. J. Health Sci.* **2012**, *4*, 34–38.

Chandra, S.; Rawat, D. S. Medicinal Plants of the Family *Caryophyllaceae*: A Review of Ethno-Medicinal Uses and Pharmacological Properties. *Integr. Med. Res.* **2015**, *4*, 123–131.

Ching-Yu, Y.; Ming-Feng, H.; Zhi-Wen, Y.; Jen-Yang, T.; Kun-Tzu, L.; Hurng-Wern, H.;

150   Green Chemistry and Biodiversity: Principles, Techniques, and Correlations

Yu-Hsuan, H.; Sheng-Yang, L.; Tzu-Fun, F.; Che-Yu, H.; Bing-Hung, C.; Hsueh-Wei, C. Concentration Effects of Grape Seed Extracts in Anti-oral Cancer Cells Involving Differential Apoptosis, Oxidative Stress, and DNA Damage. *BMC Complement. Alter. Med.* **2015**, *15*, 94.

Choi, S. K.; Zhang, X. H.; Seo, J. S. Suppression of Oxidative Stress by Grape Seed Supplementation in Rats. *Nutr. Res. Pract.* **2012**, *6* (1), 3–8.

Dumlu, M. U.; Gürkan, E. Elemental and Nutritional Analysis of *Punica granatum* from Turkey. *J. Med. Food.* **2007**, *10* (2), 392–395.

El-Beltagi, H. S.; El-Desouky, W.; Yousef, R. S. Synergistic Antioxidant Scavenging Activities of Grape Seed and Green Tea Extracts Against Oxidative Stress. *Not. Bot. Horti. Agrobo.* **2016**, *44* (2), 367–374.

El-Gengaihi, S.; Abou Baker, D. H. Grape Seeds Extract as Brain Food: A Review. *Int. J. Pharm. Clin. Res.* **2017**, *9* (1), 77–85.

Feringa, H. H. H.; Laskey, D. A.; Dickson, J. E.; Coleman, C. I. The Effect of Grape Seed Extract on Cardiovascular Risk Markers: A Meta-Analysis of Randomized Controlled Trials. *J. Am. Diet Assoc.* **2011**, *111*, 1173–1181.

Hamden, K.; Masmoudi, H.; Ellouz, F.; Elfeki, A.; Carreau, S. Protective Effects of *Peganum Harmala* Extracts on Thiourea-Induced Diseases in Adult Male Rat. *J. Env. Biol.* **2008**, *29* (1), 73–77.

Haque, N.; Sofi, G.; Ali, W.; Rashid, M.; Itrat, M. A Comprehensive Review of Phytochemical and Pharmacological Profile of Anar (*Punica Granatum Linn*): A Heaven's Fruit. *J. Ayu. Herb. Med.* **2015**, *1* (1), 22–26.

Herraiz, T.; Guillén, H.; Arán, V. J.; Salgado, A. Identification, Occurrence and Activity of Quinazoline Alkaloids in Peganum Harmala. *Food Chem. Toxicol.* **2017**, *103*, 261–269.

Hrabarova, E.; Valachova, K.; Rychly, J.; Rapta, P.; Sasinkova, V.; Malikova, M.; Soltes, L. High-molar-mass Hyaluronan Degradation by Weissberger's System: Pro- and Anti-oxidative Effects of Some Thiol Compounds. *Polym. Degrad. Stab.* **2009**, *94*, 1867–1875.

Ismail, F. A.; Abdelatif, S. H.; Abd El-Mohsen, N. R.; Zaki, S. A. The Physico-chemical Properties of Pomegranate Juice (*Punica Granatum L.*) Extracted From Two Egyptian Varieties. *World J. Dairy Food Sci.* **2014**, *9* (1), 29–35.

Jayaprakash, A. *Punica Granatum*: A Review on Phytochemicals, Antioxidant and Antimicrobial Properties. *JAIR* **2017**, *5* (9), 132–138.

Karam, M. A.; Abd-Elgawad, M. E.; Ali, R. M. Differential Gene Expression of Salt-Stressed *Peganum Harmala L. J. Gen. Eng. Biotechnol.* **2016**, *14*, 319–326.

Kim, S. Y.; Jeong, S. M.; Park, W. P.; Nam, K. C.; Ahn, D. U., Lee, S. C. Effect of Heating Conditions of Grape Seeds on the Antioxidant Activity of Grape Seed Extracts. *Food Chem.* **2006**, *97*, 472–479.

Komeili, G.; Hashemi, M.; Bameri-Niafar, M. *Evaluation of Antidiabetic and Antihyperlipidemic Effects of Peganum Harmala Seeds in Diabetic Rats*; Hindawi Publishing Corporation Cholesterol, 2016; Vol. 2016, , pp 6, Article ID 7389864.

Li, H.; Wang, X. Y.; Li, P.; Li, Y.; Wang, H. Comparative Study of Antioxidant Activity of Grape (*Vitis Vinifera*) Seed Powder Assessed by Different Methods. *J. Food Drug Anal.* **2008**, *16* (6), 1–8.

Antioxidative Properties 151

Liu, L.; Zhao, T.; Cheng, X. M.; Wang, C. H.; Wang, Z. T. Characterization and Determination of Trace Alkaloids in Seeds Extracts from Peganum Harmala Linn. Using LC–ESI–MS and HPLC. *Acta Chromatogr.* **2013,** *25* (2), 221–240.

Mandic, A. I.; Đilas, S. M.; Ćetković, G. S.; Čanadanović-Brunet, J. M.; Tumbas, V. T. Polyphenolic Composition and Antioxidant Activities of Grape Seed Extract. *Int. J. Food Prop.* **2008,** *11,* 713–726.

Mekki, L. Cytogenetic Effects of Crude Extracts of *Peganum Harmala* Seeds and Their Effects on Vicia Faba Plants. *Cytologia* **2014,** *79* (2), 161–172.

Moloudizargari, M.; Mikaili, P.; Aghajanshakeri, S.; Asghari, M. H.; Shayegh, J. Pharmacological and Therapeutic Effects of *Peganum Harmala* and its Main Alkaloids. *Pharmacogn. Rev.* **2013,** *7* (14), 199–212.

Moradi, M. T.; Karimi, A.; Rafieian-Kopaei, M.; Fotouhi, F. In Vitro Antiviral Effects of *Peganum Harmala* Seed Extract and its Total Alkaloids Against Influenza Virus. *Microb. Pathog.* **2017,** *110,* 42–49.

Nassiri-Asl, M.; Hosseinzadeh, H. Review of the Pharmacological Effects of Vitis Vinifera (Grape) and Its Bioactive Constituents: An Update. *Phytother. Res.* **2016,** *30* (9), 1392–1403.

Opara, L. U.; Al-Ani, M. R.; Al-Shuaibi, Y. S. Physico-Chemical Properties, Vitamin C Content, and Antimicrobial Properties of Pomegranate Fruit (*Punica granatum* L.). *Food Bioprocess Tech.* **2009,** *2,* 315–321.

Panth, N.; Manandhar, B.; Paudel, K. R. Anticancer Activity of *Punica Granatum* (Pomegranate): A Review. *Phytother. Res.* **2017,** *31,* 568–578.

Pirincciouglu, M.; Kizil, G.; Ceken Toptanci, B.; Ozdemir, G.; Kizil, M. Protective Effect of Öküzgözü (*Vitis vinifera* L. CV.) Seed Extract Against Hydroxyl Radical Induced DNA Damage. *Sci. Pap. Ser. B Hortic.* **2017,** *LXI,* 205–208.

Re, R.; Pellegrini, N.; Proteggente, A.; Pannala, A.; Yang, M.; Rice-Evans, C. Antioxidant Activity Applying an Improved ABTS Radical Cation Decolorization Assay. *Free Radic. Biol. Med.* **1999,** *26* (9–10), 1231–1237.

Rummun, N.; Somanah, J.; Ramsaha, S.; Bahorun, T.; Neergheen-Bhujun, V. S. Bioactivity of Nonedible Parts of Punica Granatum L.: A Potential Source of Functional Ingredients. Hindawi Publishing Corporation. *Int. J. Food Sci.* **2013,** *2013,* 1–2.

Satheesh, B.; Bhandary, K.; Suchetha, K. N.; Bhat, S. V.; Sharmila, K. P. Assessment of Antioxidant Potential of Various Extracts of *Punica Granatum Linn*: An in Vitro Study. *Int. J. Pharm. Chem. Sci.* **2013,** *2* (2), 725–729.

Shaygannia, E.; Bahmani, M.; Zamanzad, B.; Rafieian-Kopaei, M. A Review. Study on *Punica Granatum L. J.* Evid. Based Complemen. *Altern. Med.* **2016,** *21* (3), 221–227.

Soliman, A. M.; Fahmy, S. R. Protective and Curative Effects of the 15 KD Isolated Protein from the *Peganum Harmala L.* Seeds Against Carbon Tetrachloride Induced Oxidative Stress in Brain, Tests and Erythrocytes of Rats. *Eur. Rev. Med. Pharmacol. Sci.* **2011,** *15,* 888–899.

Sreemantula, S.; Nammi, S.; Kolanukonda, R.; Koppula, S.; Boini, K. M: Adaptogenic and Nootropic Activities of Aqueous Extract of *Vitis Vinifera* (Grape Seed): An Experimental Study in Rat Model. *BMC Complemen. Altern. Med.* **2005,** *5,* 1.

Stern, D. Medicinal Uses of Carnations. http://www.livestrong.com/article/70178-medicinal-uses-carnations/2016.

Suwannaphet, W.; Meeprom, A.; Yibchok-Anun, S.; Adisakwattana, S. Preventive Effect of Grape Seed Extract Against High-fructose Diet-induced Insulin Resistance and Oxidative Stress in Rats. *Food Chem. Toxicol.* **2010**, *48*, 1853–1857.

Topolska, D.; Valachova, K.; Nagy, M.; Soltes, L. Determination of Antioxidative Properties of Herbal Extracts: *Agrimonia Herba, Cynare Folium*, and *Ligustri Folium. Neuroendocrinol. Lett.* **2014**, *35* (Suppl. 2), 192–196.

Topolska, D.; Valachova, K.; Rapta, P.; Silhar, S.; Panghyova, E.; Horvath, A.; Soltes, L. Antioxidative Properties of *Sambucus Nigra* Extracts. *Chem. Pap.* **2015**, *69*, 1202–1210.

Tse, S. Y. H.; Mak, I. T.; Dickens, B. F. Antioxidative Properties of Harmane and B-Carboline Alkaloids. *Biochem. Pharmacol.* **1991**, *42*, 459–464.

Valachova, K.; Banasova, M.; Topolska, D.; Sasinkova, V.; Juranek, I.; Collins, M. N.; Soltes, L. *Carbohydr. Polym.* **2015**, *134*, 516–523.

Valachova, K.; Rapta, P.; Kogan, G.; Hrabarova, E.; Gemeiner, P.; Soltes, L. *Chem. Biodivers.* **2009**, *6*, 389–395.

Valachova, K.; Soltes, L. Effects of Biogenic Transition Metal Ions Zn(II) and Mn(II) on Hyaluronan Degradation by Action of Ascorbate Plus Cu(II) Ions. In *New Steps in Chemical and Biochemical Physics*; Eli, M., Pearce et al., Eds.; pp 149–156.

Yamaguchi, F.; Yoshimura, Y.; Nakazawa, H.; Ariga T. Free Radical Scavenging Activity of Grape Seed Extract and Antioxidants by Electron Spin Resonance Spectrometry in an $H(2)O(2)/NaOH/DMSO$ System. *J. Agric. Food Chem.* **1999**, *47* (7), 2544–2548.

Yen, C. Y.; Hou, M. F.; Yang, Z. W.; Tang, J. Y.; Li, K. T.; Huang, H. W.; Huang, Y. H.; Lee, S. Y.; Fu, T. F.; Hsieh, C. Y.; Chen, B. H.; Chang, H. W. Concentration Effects of Grape Seed Extracts in Anti-oral Cancer Cells Involving Differential Apoptosis, Oxidative Stress, and DNA Damage. BMC Complement. *Altern. Med.* **2015**, *15*, 94.

Yuan, X. Y.; Liu, W.; Hao, J. C.; Gu, W. J.; Zhao, Y. S. Topical Grape Seed Proanthocyandin Extract Reduces Sunburn Cells and Mutant P53 Positive Epidermal Cell Formation, and Prevents Depletion of Langerhans Cells in an Acute Sunburn Model. *Photomed. Laser Surg.* **2012**, *30* (1), 20–25.

Zhao, C.; Sakaguchi, T.; Fujita, K.; Ito, H.; Nishida, N.; Nagatomo, A.; Tanaka-Azuma, Y.; Katakura, Y. Pomegranate-derived Polyphenols Reduce Reactive Oxygen Species Production via SIRT3-Mediated SOD2 Activation. Hindawi Publishing Corporation. *Oxid. Med. Cell. Longev. 2016*, 1–9.

http://www.naturalmedicinalherbs.net/herbs/d/dianthus-caryophyllus=carnation.php, 2018.

CHAPTER 11

Flavonoids for Designing Metal Nanoparticles and Their Applications

DIVYA MANDIAL, RAJPREET KAUR, and POONAM KHULLAR[*]

Department of Chemistry, B.B.K. D.A.V. College for Women, Amritsar 143005, Punjab, India

[*]Corresponding author. E-mail: virgo16sep2005@gmail.com

ABSTRACT

Recent developments show the high potential of microorganisms and biological products for designing nanometals. Plant extracts including enzymes, proteins, amino acids, vitamins, polysaccharides, and organic acids have been reported to provide environmentally sociable protocols in material synthesis. Flavonoids act as natural reducing and capping agent and therefore have emerged as the promising candidate in the field of nanobiotechnology. The present study reports the efficiency of flavonoids including hesperidin, naringin diosmin, hesperetin, rutin, narengenin, quercetin for nanometals synthesis. Metals including gold, silver, copper, and palladium have been successfully transformed into nanostructures and well-characterized. The antimicrobial and antioxidant property of flavonoid conjugated nanostructure displays the versatility of fields applicable.

11.1 INTRODUCTION

Flavonoids are the diverse group of polyphenols found in almost all fruits and vegetables along with tea and wine. They are responsible for color and taste of food and also play a significant role in prevention of fat oxidation and protection of vitamins and enzymes. The increased research on flavonoids is on the account of their versatile health benefits for mankind.

Flavonoids are also called "nature's biological response modifiers" because of their antioxidative, anti-inflammatory, antiallergenic, antiviral, anticancer free radical scavenging, coronary heart disease prevention, and hepatoprotective properties. Also, a number of studies have suggested the protective action of flavonoids against many infectious (bacterial and viral diseases) and degenerative diseases such as cardiovascular diseases, cancers, and other age-related diseases.[1,2-4] The basic structure of flavonoid consists of 15-carbon skeleton consisting of two benzene rings (A and B) linked via a heterocyclic pyrane ring (C) as shown in Figure 11.1. They can be categorized into a variety of classes including flavones (e.g., flavone, apigenin, and luteolin), flavonols (e.g., quercetin, kaempferol, myricetin, and fisetin), flavanones (e.g., flavanone, hesperetin, and naringenin), isoflavones (daizein, genistein, and glycitein), flavan-3-ols (epicatechingallate and catechins), and anthocynins (pelargonidin, delphinidin, peonidin, petunidin, and malvidin). The various classes of flavonoids differ in the level of oxidation and pattern of substitution of the C ring, while individual compounds within a class differ in the pattern of substitution of the A and B rings.[5] The presence of polyhydroxy group on flavonoids make them valid for biomimetic synthesis of metal nanoparticles. In such biogenic synthesis, flavonoids not only act as reducing agent but also as stabilizing agent. The coordination reduces the electron density on flavonoid, and it acts as effective reducing agent to convert metal into its corresponding nanoparticles. The concentration of plant extract and metal salt along with other factors such as pH, temperature, and reaction time controls the size and morphology of NPs. Increasing demand for biologically synthesized NPs using plant extracts, biodegradable polymers, bacteria, enzymes, and fungi have replaced hazardous chemicals and an alternate form of heating, microwave irradiation. Flavonoid based NPs, besides their antimicrobial properties, have also displayed suitable applications in medicine, disease diagnostic, drug delivery systems, and biogenic sensors. This chapter deals with flavonoids which have been discovered in last few years for designing metal nanoparticles.

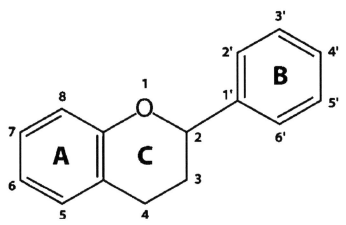

FIGURE 11.1 Basic flavonoid structure.
Source: Reprinted with permission from Ref. [5]. https://www.hindawi.com/journals/tswj/2013/162750/

11.2 FLAVONOIDS FOR DESIGNING NANOMETALS

Hesperidin, naringin, and diosmin have been recently explored to design silver nanoparticles.[6] Hesperidin and naringin are flavanone glycosides which naturally occurs in citrus fruits while diosmin naturally occur in *Mentha pulegium*. In UV spectrophotometer, the single absorption peak at around 447, 438, and 492 nm confirmed the synthesis of pure AgNPs. The presence of double bond in the pyran ring of diosmin affects the reducing ability of diosmin in comparison to hesperidin and naringin. The XDR analysis confirms the polydispersity of synthesized AgNPs which is in the order—hesperidin > naringin > diosmin. The intensity peak at 2Θ values of 35.2, 29.2, 29.5 for hesperidin, naringin, and diosmin, respectively confirms the crystalline nature of synthesized AgNPs. FTIR spectrum depicts the major shift in band intensities which were observed at 3600–3200, 3000–2500, and 1600–1400 cm^{-1}. A weak signal is detected in the range of 3000–3500 cm^{-1} which indicates the decreased concentration of OH groups in the sample after synthesis of AgNPs. Transmission electron microscopy determines the size of AgNPs synthesized from hesperidin, naringin, and diosmin in the range of approximately 5–50, 5–40, and 20–80 nm which were oval, polydispersed, and hexagonal in shape respectively as shown in Figure 11.2.

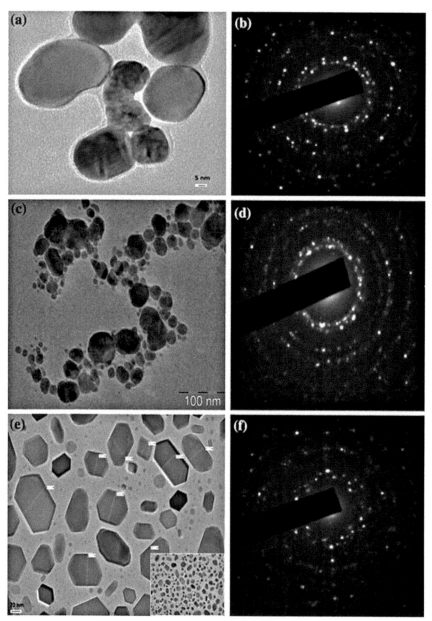

FIGURE 11.2 Transmission electron micrographs of silver nanoparticles synthesized by hesperidin (a), naringin (c), and diosmin (e), and their respective electron diffraction pattern (b, d, f).
Source: Reprinted with permission from Ref. [6]. © 2016 Springer Nature.

The antimicrobial activity of synthesized NPs when studied using disc diffusion method showed the antimicrobial activity against *Escherichia coli, Pseudomonas putida, Staphylococcus aureus* as mentioned in the table. AgNPs are reported for the formation of pits on bacterial cell wall and alteration of cell wall permeability, inhibition of transduction, inhibition of respiratory enzymes function due to free radical formation and inactivation of several enzymes having thiol group.[7] Flavonoids acts as capping agents to synthesized AgNPs, therefore, affecting the antimicrobial activity of AgNPs as shown in Figure 11.3. The cytotoxicity of synthesized by NPs was investigated in HL-60 cells by MTT assay at different concentrations LC_{50} values of 13.7, 4.85, 19.55 µm for hesperidin, naringin, and diosmin showed the high toxic potential of naringin in comparison to other two. This may be attributed to the smaller size and monodispersed AgNPs of Naringin as compared to large and polydispersed AgNPs. Thus, the AgNPs designed in this manner have potential to be used as a drug for cancer patients.

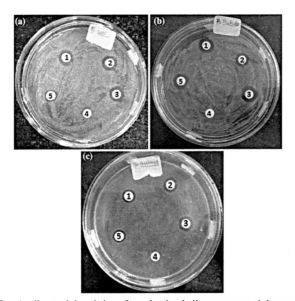

FIGURE 11.3 Antibacterial activity of synthesized silver nanoparticles against common pathogen, (a) *Escherichia coli*, (b) *Pseudomonas putida*, and (c) *Staphylococcus aureus*, respectively. (1) Hesperidin synthesized AgNP, (2) Naringin synthesized AgNP, (3) Diosmin synthesized AgNP, (4) negative control (autoclaved distilled water), and (5) positive control (1 mM $AgNO_3$).
Source: Reproduced with permission from Ref. [6]. © 2016 Springer Nature.

Citrus peel extracts from oranges, tangerines, and lemon are suggested to have high potential for synthesizing AgNPs without addition of any external surfactants or capping agents.[8] The UV spectrum showed the localized surface Plasmon resonance band at 445 nm and 423 nm, respectively for AgNPs synthesized from orange peel and tangerine peel extracts. The formation of ultra-small NPs/silver clusters in case of lemon peel extracts showed no localized surface Plasmon band. High-resolution TEM images of AgNPs synthesized showed the average sizes of 31 ± 18.3, 29.8 ± 18.7, and 18.5 ± 11.6 nm for orange tangerine and lemon peel extract synthesized AgNPs. The self-assembled AgNPs plasmonic substrates can also be prepared by one or two depositions of AgNPs which forms aggregates and thus forming rich areas of AgNPs even for lemon peel extract synthesized AgNPs as shown in SEM images. EDS elemental analysis showed higher percentage of Ag recorded for tangerine peel extract synthesized AgNPs due to larger AgNPs aggregate in comparison to other two substrates. Surface enhance Raman scattering (SERS) was observed in all AgNPs aggregate substrates which was evaluated by using 4-amino benzenethiol (4-ABT) rhodamine 6G and methylene blue as Raman probes. The results proved that citrus peel extracts NPs can be efficiently employed in SERS like any other synthetic AgNPs. The SERS intensities for AgNPs using 4ABT probe were higher as compared to dye probes.

AuNPs can also be prepared using citrus flavonoids as reductants. Hesperidin, hesperetin, rutin, narengenin, quercetin, and diosmin have been successfully reported to be responsible for both reduction and stabilization of AuNPs in alkali medium.[9] Hesperetin and its glycoside from hesperidin, naturally occurs in peels of lemon and oranges and displays anti-inflammatory, antiantherogenic, antitumor, antihypertensive, antioxidants, antifungal, antiallergic, antidepressive, neuro-protective, and improves memory and learning, therefore a multifunctional flavanoid.[10] The appearance of localized surface Plasmon resonance band (LSPR) at 520–550 nm UV-visible spectra showed diosmin and hesperetin to be effectively producing homogeneous AuNPs, whereas quercetin and rutin were able to reduce Gold ions without stabilizing them. Hesperidin could not synthesize AuNPs. The size of HtAuNPs as measured from dynamic light scattering technique (DLS) showed the particles up to 22 nm whereas size, as measured from transmission electron microscope (TEM), showed the size of 15 nm. This may be due to TEM recording only the Au core whereas DLS measuring the surface chemisorption of Hesperetin in

Flavonoids for Designing Metal Nanoparticles and Their Applications 159

addition to Au core. The crystalline nature of HtAuNPs was confirmed by selected area electron diffraction (SAED) analysis. The diffraction rings were assigned to (111), (200), (220), (311), and (222) planes of face-centered cubic (FCC). The X-ray photoelectron spectroscopy (XPS) displayed the binding energy of Au 4f 84.38 and 88.07 eV with spin–orbit splitting of 3.68 eV and area ratio of 0.7 which is in agreement with literature values of pure metallic gold.[11] Also, there are minor contributions of Au^{1+} and Au^{3+}. The data favors O–Au binding. The spectrum also shows the appearance of C, O, Na, Cl, Si peaks in addition to Au peaks. The C 1s envelope of hesperetin and hesperetin mediated Au particles as deposited on Si.

Bioinspired synthesis of AuNPs can be done by using naringin, a flavonoid which acts as both reducing and capping agent for AuNPs under mild conditions.[12] In UV Visible spectroscopy, the peak at 336 and 238 nm decreased on adding $HAuCl_4$ as both phenoxide and condensed rings are involved in reduction of Au^{3+} to Au^0 with appearance of SPR band at 525 nm . As the reduction proceeds, $AuCl_4^-$ ions bind to side chains of naringin through ligand to metal charge transfer complex. TEM analysis showed the naringin coated AuNPs with size range of 20–25 nm. FTIR studies revealed the merging of bands at 1581, 1520, and 1450 cm^{-1} in N due to aromatic C=C into a single band at 1540 cm^{-1} showing the presence of naringin in AuNPs. The band of phenolic C–OH group shifted from 1345 to 1378 cm^{-1} with a band at 1294and 1203 cm^{-1} merging into a single band at 1224 cm^{-1}. This confirms the participation of OH group in AuNPs synthesis. Five bands due to glucose C–OH stretching merged into a band at 1163 and 1050 cm^{-1} indicating the presence of glucose in naringin coated AuNPs. The XRD patterns confirmed the FCC geometry of nanocrystal with predominant growth at (111) crystal planes. In gel electrophoresis, the negatively charged AuNPs showed movement toward positive terminal which may be attributed to the solvation of OH group of naringin coated AuNPs. The binding energy calculations displayed that both functional group $[C_6H_5O]^-$ and $[C_{10}H_7O_2]$ interacts strongly with Au clusters with B.E values of 1.88 eV and 2.15 eV, respectively. The metal ions involving Al^{3+}, Fe^{3+}, Cr^{3+}, Ba^{2+}, Mg^{2+}, Sr^{2+}, Cd^{2+}, Co^{2+}, Mg^{2+}, Mn^{2+}, and Ag^+ were detected at concentration of 20 mM for the application of naringin coated AuNPs as calorimetric sensor. The remarkable change in intensity and wavelength was observed from 25 to 545 nm and 1.03 to 0.78 in case of Al^{3+} in comparison to other metals which showed high selectivity and sensitivity of naringin coated AuNPs toward Al^{3+}. The average size of AuNPs also increased from 23 to 35 nm in presence of Al^{3+}. This provides a simple, rapid, and cost-effective method for sensing Al^{3+} in real water samples with detection limit of 0.1 mg mL^{-1}. The naringin coated AuNPs with minimum concentration of 25 mg/mL showed least damage to cell walls, therefore, may be used as drug release vehicles in systemic circulation. Furthermore, naringin coated AuNPs are cytotoxic to cancer cells in dose-dependent manner in the conc. range of 30–100 µg/mL. Thus, naringin offers nontoxic and biofriendly AuNPs for biomedical and sensing applications.

Quercetin, a flavonoid is one of the natural plant pigments that have been explored to synthesize AgNPs, CuNPs, and bi-metallic NPs through reverse micelle particle synthesis. The method significantly enhances the particle stability up to years for its practical applications.[13] The UV visible spectrum showed the elimination of quercetin band and appearance of new absorption bond at wavelength of 400–410 nm with addition of $AgNO_3$ and a band at wavelength 425–440 nm with addition of $CuNO_3$ solution. It was recorded that for a particular concentration of Ag and Cu, the final intensity of bands increased with increased concentration of quercetin. The analyses of absorption spectrum of Ag NPs in terms of mie theory revealed that with increase of particle diameter the absorbance also increases with simultaneous decrease in half-width of Plasmon band. And for constant particle size of 1 nm, the increase in Ag conc. increased the absorbance with half-width remain unchanged and equal to 410 and 140 nm.

AgNPs and AuNPs can also be synthesized using antioxidants from blackberry, blueberry, pomegranate, and turmeric. The fruit extracts of these provide greener, cost-effective synthesis of AuNPs and AgNPs with their capping and reducing properties.[14] Different tea and coffee varieties have also been reported for the synthesis of metal NPs.[15] The plant extracts are rich in flavonoids: catechin, quercetin, peonidin, malvidin, pelargonidin, fisetin, and myricetin. These flavonoids along with polyphenols, anthocyanins, and antioxidants are collectively responsible for regulating the nano-particles size and uniform shapes. The XRD patterns and AgNPs as synthesized using blackberry, blueberry, pomegranate, and turmeric extracts showed peak at 2Θ value of 38.2°, 44.4°, and 64.66° corresponding to (111), (200) planes of Ag, respectively. Similarly, peaks at 2Θ values of 38.17°, 44.37°, 64.55°, and 77.54° correspond to (111), (200), and (222) planes of Au, respectively. The TEM micrographs of AgNPs and

Au NPs as synthesized with blueberry extracts displayed the average size of 50 to 150 nm (spherically and triangular shaped) and 200 nm (spherical shaped), respectively as shown in Figure 11.4a, b. Likewise, Ag and Au NPs as synthesized with blackberry extracts displayed the average size of 25 to 150 nm and 100 nm (oblong shaped) respectively as shown in Figure 11.4c, d. Likewise, Ag and Au NPs as synthesized with pomegranate extract displayed the average size of 5 to 50 nm (spherical shaped) and 400 nm (bigger spherical particles), respectively as shown in Figure 11.5a, b. Likewise, Ag and Au NPs as synthesized using turmeric extracts displayed the average size of 100 nm (large oblong spheres), 5 to 10 nm (small spherical spheres) and 5 to 60 nm (spherical shaped) respectively as shown in Figure 11.5c, d. The UV Visible spectra of Ag NPs displayed a broad plasma resonance peak at 420–460 nm with an additional peak around 550–650 nm corresponding to dipole and quadrupole modes of Ag due to different shapes and sizes whereas AuNPs displayed a plasma resonance peak around 560 and 650–850 nm. These NPs with antioxidant coatings have relevant potential to be used in delivery of useful oxidants, cancer, and chemo preventive agents based on curcuminoids.

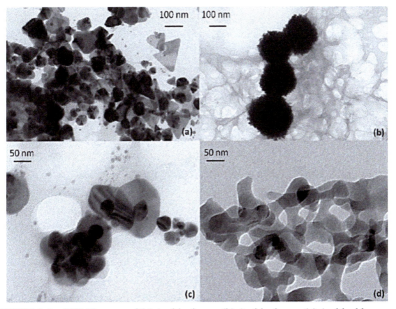

FIGURE 11.4 TEM images of (a) Ag blueberry, (b) Au blueberry, (c) Ag blackberry, and (d) Au blackberry.
Source: Reprinted with permission from Ref. [14]. © 2014 American Chemical Society.

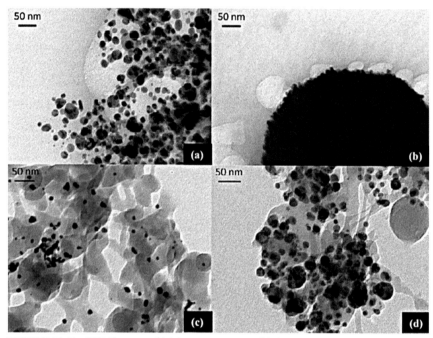

FIGURE 11.5 TEM images of (a) Ag pomegranate, (b) Au pomegranate, (c) Ag turmeric, and (d) Au turmeric.
Source: Reprinted with permission from Ref. [14]. © 2014 American Chemical Society.

Plant leaf extracts of tamarind, syzgium, cumini, onion, basil, and banana peel have been reported for effective synthesis of pure metal NPs.[16–20] Flavone enriched *Ficus benghalensis* leaf extracts have also been reported to synthesize AgNPs with 5 min of reaction time and under mild conditions.[21] The UV visible spectrum showed the appearance of SPR band at 410 nm with remarkable increase in intensity peak with increased reaction time. The X-ray diffraction pattern showed the characteristic peak at 2Θ 80.02° with unit cell of FCC structure. The nano regime as analyzed by TEM showed the spherical morphology quoted with flavone with average size of 16 nm approximately. The energy dispersive X-ray (EDX) spherometer analyses confirmed the presence of Ag from the major emission energies of Ag that were displayed. The proteins and enzymes also play a characteristic role in the reduction of meta-ions. The Ag nanoparticles showed efficient bactericidal properties as compared to Ag salts on account of higher surface to volume ratio. It was observed that the

Flavonoids for Designing Metal Nanoparticles and Their Applications   163

concentration of AgNPs of 45 µg/mL acted as an efficient bactericide as compared to the minimum concentration of 25 µg/mL against *E. coli.*

Black tea leaves possess considerable potential to synthesize AuNPs and AgNPs. They are rich in polyphenolic compounds especially flavonoids.[22] Black tea leaf extract in the form of tea leaf broth, ethyl acetate extract, and CH_2Cl_2 extract can be used for biogenic reduction of Ag^{1+} and Au^{3+} ions. The UV spectrum as recorded for AgNPs using tea leaf broth and ethyl acetate extract showed the appearance of peak at 438 nm and 460 nm, respectively which are assigned to SPR band of Ag NPs. The UV spectrum of Au NPs synthesized using tea leaf broth and ethyl acetate extract showed the appearance of peak at 542 nm and 557 nm, respectively which are assigned to SPR band of AuNPs. Both AgNPs and AuNPs showed the initial increase of absorbance, as well as peak wavelength, followed the completion of reaction leading to saturation after first 200 min and 300 min, respectively. Also, it was noticed that the reduction of Au was kinetically more facile as compare to Ag under similar condition of extract solution and metal ion concentration. While no NPs formation was observed for CH_2Cl_2 extract under similar condition due to lack of polyphenol in comparison to other extract. The TEM micrographs confirmed the formation of spheroidal NPs, nanoprisms, and nanorods with size range of 20 nm approximately in case of AgNPs while AuNPs were more or less spherical with formation of nanoprisms and dumbell like particles. The FTIR absorption spectrum showed a band at 1226 cm^{-1} which is attributed to CO group of polyols .The disappearance of this band after bioreduction may be due to reduction of Ag^{1+} and Au^{3+} by polyols. They themselves get oxidized to α, β unsaturated carbonyl group leading to a broad peak at 1650 and 1641 cm^{-1} for reduction of Ag^{1+} and Au^{3+}, respectively. The reduction peak for tea leaf broth and ethyl acetate extract appeared at −0.44 V as recorded by cyclic voltmeter but the peak for ethyl acetate extract was more sharp compared to tea leaf broth. The efficiency of naturally occurring hydroxyl flavonoid, quercetin was analyzed for synthesis of Au and Ag NPs and it was seen that efficiency of quercetin toward Ag ions was too low to cause any reduction. On the other hand, quercetin played a major role in the reduction to Au^{3+} to Au^0.

Cycas leaves have been found to be rich in Amenti flavones and Hinokiflavone as characteristic bioflavonyls.[23–26] Therefore, taken as potential candidate plant specimen for synthesis of metal and oxide NPs.[27] A green and low-cost synthesis of AgNPs has been reported using Cycas leaf extracts. The TEM micrographs showed the spherical shaped Ag NPs with a diameter of 2–6 nm. The FCC Structure of Ag NPs as recorded using XRD and SAED confirmed the nanocrystalline nature of AgNPs. The UV-Visible spectra showed silver surface Plasmon resonance band at 449 nm, the intensity of which increased steadily with time. The broadened Plasmon band is attributed to size distribution of particles. The reduction process by cycas leaf is accomplished due to redox activities of ascorbic acid/dehydroascorbic acid, amenti/hinoki flavones, and ascorbates/glutathiones/metallothioneins, thus, a better candidate to synthesize AgNPs.

The nanoformulated insecticides and antimicrobials in the form of AgNPs can be synthesized using *Bauhinia acuminata* leaf extract which are suggested to have high bioactivity against pathogens and vectors.[28] *Bauhinia acuminata* possesses antioxidant, antibacterial, antifungal, and hemolytic activities. Its leaves and stem are rich in phenolic compounds, flavonoids which are responsible for reduction and capping of Nanosilver. The UV Visible spectrum showed the peak at 441.5 nm which is assigned to SPR band of AgNPs. The low transmittance band at 466 cm^{-1} corresponding to Ag was observed in FTIR. The XRD pattern confirmed the FCC crystal structure of Ag with the mean crystalline size of 26.5 nm which was in agreement with mean value of TEM analysis. *Bauhinia acuminata* leaf extract synthesized AgNPs showed efficient antibacterial potential in *E. coli* and *S. aureus* at 900 and 750 µg concentration.[29] *Bauhinia acuminata* fabricated AgNPs showed significant antibacterial activity on *S. aureus*, *S. pneumonia*, *E. coli*, *P. aeruginosa*, and *S. dysenteriae* at 60 µg/mL concentration. The antifungal potential of *B. acuminata* exhibited excellent antifungal activity against *A. fumigatus*, *A. niger*, and *C. altricans* at the effective concentration of 60 µg/mL. This concentration of AgNPs interrupts the intracellular communication of fungi thereby reducing the lifespan of fungi and significantly arresting the growth of fungal strains. Furthermore, the *B. acuminate* synthesized AgNPs demonstrated eight times higher mortality rate than *B. acuminate* of extract treatment itself. The LC$_{50}$ values of 24.59, 20.19, 30.19 µg/mL on larva of *An. stephensi*, *Ae. aegypti*, and *Cx. quinquefasciatus*, as recorded with AgNPs, were much lower than the extract of *B. acuminata*. AgNPs penetrates the Cell cytoplasm and binds to sulfur-containing amino acids leading to formation of stable metal-protein complex which is toxic for mosquito larva. AgNPs also interfere with the electron transport chain in

Flavonoids for Designing Metal Nanoparticles and Their Applications 165

mitochondrion; therefore, suppressing the ATP synthesis in cell and leading to final breakdown of metabolic activities in the cell. AgNPs also causes chromosomal aberrations and reduce life span of cell. It also inactivates the peroxidase enzyme in midgut epithelial membrane leading to cell death.[30–34] Therefore the AgNPs synthesized from *B. acuminate* finds relevant application in antimicrobial food packaging as well as foliar spray to control plant pathogens in the field and to design new fungicidal and larvicial formulations.

11.3 CONCLUSION

The biosynthesis using flavonoids has stimulated eco-friendly techniques for designing well-characterized nanoparticles. The resulted UV-Visible studies, IR spectroscopy, XRD patterns, XPS analysis, TEM, SEM, and SERS studies were in accordance with literature listed values of nanocrystals. The growth factors such as pH, reaction time, structure, and concentration of flavonoid and concentration of metals help in directing the shape, size, and quantity of nanoparticles. Hesperidin, naringin, and diosmin conjugated Ag nanoparticles have efficient antimicrobial activity against *E. coli*, *P. putida*, *S. aureus*. Naringin coated AgNPs shows considerable cytotoxicity against human promyelocytic lymphoma cells, therefore, can be used as drug against cancerous cells. Naringin-coated AuNPs have been discovered for designing cost effective and simple technique to detect Al^{3+} ions from real samples with the detection limit of 0.1 mg/mL. Also, these NPs can be tagged as suitable drug releasing vehicles due to their biocompatibility with normal and cancerous cells. The AgNPs designed from leaf extracts of *Ficus benghalensis* has shown efficient antibacterial action at the concentration of 45 µg/mL as compared to its silver salts. The AgNPs designed from leaf extracts of *B. acuminata* possess high antibacterial activity against *S. aureus*, *S. pneumonia*, *E. coli*, *P. aeruginosa*, and *S. dysenteriae* at 60 µg/mL concentration, antifungal activity against *A. fumigatus*, *A. niger*, and *C. altricans* and larvicidal activity against larva of *An. stephensi*, *Ae. aegypti*, and *Cx. quinquefasciatus*. With emerging focus on flavonoids conjugated metal nanoparticles, a lot many applications have come in interest and yet many more are to be uncovered.

166 Green Chemistry and Biodiversity: Principles, Techniques, and Correlations

KEYWORDS

- applicability
- characterization
- flavonoids
- nanometals
- nanoparticles

REFERENCES

1. Pandey, A. K. *Natl. Acad. Sci. Lett.* **2007,** *30* (11–12), 383–386.
2. Kumar, S.; Gupta, A.; Pandey, A. K. *ISRN Pharmacol.* **2013,** *1,* 8.
3. Cook, N. C.; Samman, S. J. *Nutr. Biochem.* **1996,** *7* (2), 66–76.
4. Rice-Evans, C. A.; Miller, N. J.; Bolwell, P. G.; Broamley, P. M.; Pridham, J. B. *J. Free Rad. Res.* **1995,** *22* (4), 375–383.
5. Kumar, S.; Pandey, A. K. Chemistry and Biological Activities of Flavonoids: An Overview. *Sci. World J.* **2013,** *1,* 16.
6. Chandrashekhar, B.; Sahu, N.; Soni, D.; Satpute, D. B.; Saravanadevi, S.; Sarangi, B. K.; Pandey, R. A. Synthesis and characterization of silver nanoparticles using Cynodon dactylon leaves and assessment of their antibacterial activity. *Bioprocess. Biosyst. Eng.* **2016,** *6* (3), 173–181.
7. Umoren, S. A.; Obot, I. B.; Gasem, Z. M. *J. Mater. Environ. Sci.* **2014,** *5,* 907–914.
8. Santos, E. B.; Madalossi, N. V.; Sigoli, F. A.; Mazali, I. O. *New J. Chem.* **2015,** *39,* 2839–2846.
9. Sierra, J. A.; Vanoni, C. R.; Tumelero, M. A.; Cid, C. C.; Faccio, R.; Franceschini, D. F.; Pasa, A. A. *New J. Chem.* **2016,** *40,* 1420–1429.
10. Roohbakhsh; Parhiz, H.; Soltani, F.; Rezaee, R.; Iranshahi, M. *Life Sci.* **2014,** *113,* 1–6.
11. Herranz, T.; Deng, X.; Cabot, A.; Alivisatos, P.; Liu, Z.; Soler-Illia, G.; Salmeron, M. *Catal. Today* **2009,** *143,* 158–166.
12. Singh, B.; Rani, M.; Singh, J.; Moudgil, L.; Sharma, P.; Kumar, S.; Tripathi, S. K.; Singh, G.; Kaura. A. *RSC Adv.* **2016,** *6,* 79470–79484.
13. Egorova, E. M.; Revina, A. A. *Coll. Surf. A Physicochem. Eng. Aspects* **2000,** *168,* 87–96.
14. Nadagouda, M. N.; Iyanna, N.; Lalley, J.; Han, C.; Dionysiou, D. D.; Varma, R. S. Synthesis of Silver and Gold Nanoparticles Using Antioxidants from Blackberry, Blueberry, Pomegranate, and Turmeric Extracts. *ACS Sustain. Chem. Eng.* **2014,** *2,* 1717–1723.
15. Nadagouda, M. N.; Varma, R. S. *Green Chem.* **2008,** *10,* 859–862.
16. Ankamwar, B,; Chaudhary, M.; Sastry, M. *Nanometal Chem.* **2005,** *35,* 19–26.
17. Kumar, V.; Yadav, S.C.; Yadav, S. K. *J. Chem. Technol. Biot.* **2010,** *85,* 1301–9.

18. Saxena, A.; Tripathi, R. M.; Singh, R. P. *Dig. J. Nanomater. Bios.* **2010**, *5*, 427–32.
19. Ahmad, N.; Sharma, S.; Alam, M. K.; Singh, V. N.; Shamsi, S. F.; Mehta, B. R. et al. *Coll. Surf. B* **2010**, *81*, 81–86.
20. Bankara, A.; Joshi, B.; Ravi, A.; Zinjardea, S. *Coll. Surf. B.* **2010**, *80*, 45–50.
21. Saxena, A.; Tripathi, R. M.; Zafar, F.; Singh, P. *Mater. Lett.* **2012**, *67*, 91–94.
22. Begum, N. A.; Mondal, S.; Basu, S.; Laskar, R. A.; Mandal, D. *Coll. Surf. B Biointerf.* **2009**, *71*, 113–118.
23. Harborne, J. B., Mabry, T. J., Mabry, H., Eds.; Chapman and Hall: London, 1975; p 693.
24. Harborne, J. B. Chapman and Hall: London, 1973; p 52.
25. Mabry, T. J.; Markham, K. R.; Thomas, M. B. *The Systematic Identification of Flavonoids;* Springer-Verlag: New York, 1970; p 215.
26. Goodwin, T. W., Ed.; *Chemistry and Biochemistry of Plant Pigments;* Academic Press: New York, 1976; p 736.
27. Jha, A. K.; Prasad, K. *Int. J. Green Nanotechnol. Phys. Chem.* **2010**, *1*, 110–117.
28. Alharbi, N. S.; Govindrajan, M.; Kadaikunnan, S.; Khaled, J. M.; Almanaa, T. N.; Alyahya, S. A.; Gopinath, K.; Sudha, A. *J. Trace Elem. Med. Biol.* **2018**, *50*, 146–153.
29. Sebastian, D.; A. T. *Int. J. Sci. Res.* **2017**, *6*, 50–55.
30. Benelli, G. *Acta Trop.* **2018**, *178*, 73–80.
31. Nalini, M.; Lena, M.; Sumathi, P.; Sundaravadivelan, C. *J. Basic Appl. Sci.* **2017**, *4*, 212–218.
32. Benelli, G. *Parasitol. Res.* **2016**, *115*, 23–34.
33. Benelli, G. *Parasitol. Res.* **2015**, *114*, 2801–2805.
34. Benelli, G. *Environ. Sci. Poll. Res.* **2018**, *25*, 12329–12341.

CHAPTER 12

pH and Temperature Factor Affecting Curcumin Properties and Its Bioapplicability

RAJPREET KAUR, DIVYA MANDIAL, LAVANYA TANDON, and POONAM KHULLAR[*]

Department of Chemistry, B.B.K. D.A.V. College for Women, Amritsar 143005, Punjab, India

[*]Corresponding author. E-mail: virgo16sep2005@gmail.com

ABSTRACT

Curcumin, a natural polyphenol, shows some clinical activities. Poor solubility and less bioavailability, encapsulation method, nanonizations, micelle formation, and so on of hydrophobic drug, curcumin, is found to be very effective to increase the solubility of curcumin also increased its bioavailability. Water soluble polymer biomaterials combined with curcumin in form of mpeg-micelles, silver hydrogels, MMT formulation, polymer block have increased the thermal stability, solubility, and loading capacity of drug and efficiency of drug release encapsulation. They can be used as nano-carriers for hydrophobic drug in target drug delivery. These nanocarriers also increase the stability on surface as well as interior of nanoparticles. Release rate of drug is determined by the properties of polymer matrix of nanocarriers.

12.1 INTRODUCTION

Curcumin, widely known as turmeric, is a natural polyphenol obtained from rhizome of *Curcuma longa*,[1] a perennial herb. Curcumin (diferuloylmethane), having low molecular weight (F.W. 368.37), is available in 31 species of curcuma plants, a member of the ginger family *Zingiberaceae*.

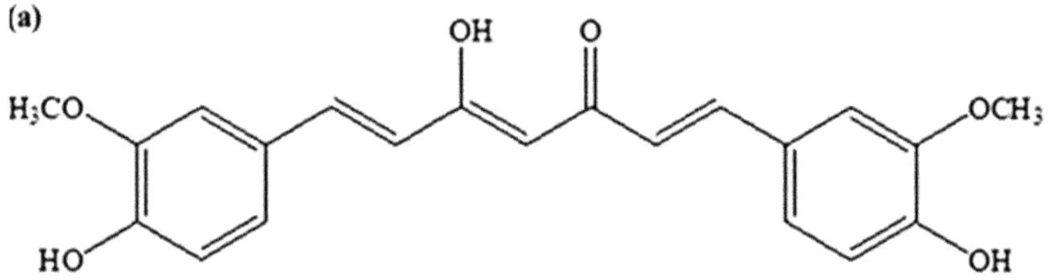

FIGURE 12.1 Structure of curcumin (enolic form).
Source: Reprinted with permission from Ref. [45]. © 2016 Korean Institute of Chemical Engineers, Seoul, Korea.

Curcumin has medicinal properties like anti-inflammatory,[2] antibacterial,[3] antifungal,[4] anticancer,[5] antispasmodic,[6] antioxidant,[7] antiamoebic,[8] anti HIV,[8] antidiabetic,[9] and antifertility.[10] Besides these properties, the use of curcumin is very limited because of its low solubility and low bioavailability, fast metabolic rate, rapid degradation, and fast systemic elimination.[11] Various necessary systems have been elaborated to increase the solubility in aqueous medium and for drug delivery to the targeted site with increased efficiency and reduced side effects on healthy cells and organs. Solid dispersion technique is most preferred to improve the solubility of curcumin in aqueous medium. It fastens with inert transporter or matrix in solid state by melting, melting solvent method, or dissolution in solvent. Various water soluble polymers, like PEG, PVP, PVA have been used as dispersion matrix. Curcumin, a hydrophobic drug could enhance its use clinically by encapsulated with hydrogels, micro and nanoparticles, polymer drug conjugates, cyclodextrins, and micelles. Their diameter ranges from micrometers to nanometers. These nanocarriers have large bio applicability due to their large specific surface area. Also, we can change the morphology and size of nanoparticles by changing the concentration in the gel formulations. In addition to this, the release rate of drug in intravenous fluid depends upon the properties of matrix prepared by encapsulation method. When the encapsulated drug is inserted in the fluid media, polymer dissolves and the drug is released as colloidal particles.[12] Cucumin drug, with medicinal properties[13] conjugated with micelles, can locate itself into the tumor without effecting natural tissues. Keto-enol tautomerism is basically observed in curcumin. In neutral and acidic medium, keto form of curcumin is prominent shows influential hydrogen patron.[15–16] Some micelles like PDEA blocks are amphoteric in nature by protonation and deprotonation. These pH sensitive micelles are attracted by tumor, inflammatory tissue and so on, due to difference between tumor tissues and normal tissues. So, they are also named as pH –triggered drug release in tumor tissues. Reduction in particles size from micronization to nanonization improves the dissolution rate and oral bioavailability. A highly stabilized dried nanocurcumin with mean particle size of 80 nm formulated with remarkable reproducibility and easy storage. Oral administration to Tg2576 AD model mice even at a low dose that is, 23 mg/kg per week, resulted in significant improvements over placebo control in working and cue memory. Pharmacokinetic studies showed that the nanoparticle formulation significantly improved curcumin bioavailability, with a much greater plasma concentration and 6-fold higher AUC and MRT in brain. This novel nanocurcumin may have great potential for Alzheimer disease therapy and can be a reference for future studies on formulation of drugs as nanoparticles. MPEG-PCL micelles to deliver curcumin, developing an intravenously injectable aqueous formulation for curcumin. This formulates its antiangiogenesis and anticancer study in vitro and in vivo. MPEG-PCL micelles have potential application in drug delivery systems and are already used to deliver some drugs.[17]

12.2 DISCUSSION

Curcumin has poor solubility in water that result in poor bio-availability. To improve the solubility of curcumin in water solid dispersion method[14] like hot melt method and solvent evaporation were used. PEG-4000, PEG-6000, Tween 80, and PVP K 30 were chosen as carriers for solvent evaporation method. Solubility of curcumin in solid dispersion increased about 100-folds with PEG-6000 by hot melt method as compared with hot melt method, solubility of curcumin decreases with PVP K 30 by solvent evaporation method. Further evaluation of studies like SEM, TLC, IR, X-ray analysis,[18] pure curcumin, curcumin solid dispersions by hot melt method (SDHM) and by solvent evaporation method in vitro dissolution profile and release of drug in medium was 2.6%, 10.03%, and 8.5%, respectively. It has been deduced from the TLC studies that curcumin (Rf, 0.96), demethoxy curcumin (Rf, 0.94), and bis-methoxy curcumin

(Rf, 0.88 to 0.9) did not show any interaction between the drug and carrier.[18]. In Figure 12.2, poor solubility and stability of curcumin (bottle 1), bare ortho-carboxymethylchitosan NPs (o-cmc NPs) (bottle b), and in (bottle c) curcumin o-cmc NPs have been shown. It is clear from Figure 12.1 and above statement that curcumin needs platform to make itself dissolvable in aqueous medium.

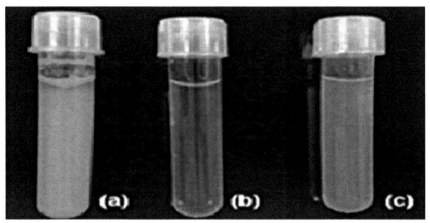

FIGURE 12.2 **(See color insert.)** Solubility of curcumin (a) curcumin in water, (b) O-CMC NPs, and (c) curcumin-OCMC NPs.
Source: Reprinted with permission from Ref. [45]. © 2016 Korean Institute of Chemical Engineers, Seoul, Korea.

Solubility of curcumin with reduced side effects could be enhanced by its physically encapsulation into the hydrophobic cores of polymeric micelles PCL-PDEA-PSBMA, INVITE micelles, MMT formulation, and so on.

12.2.1 EFFECT OF pH

pH affects the keto-enol tautomerism. pH less than 1, curcumin occurs in protonated form, and at high pH, deprotonation form of curcumin exists. Deportation helps curcumin to soluble in aqueous medium. However, under alkaline conditions by Tonneson and Wang et al. and at high temperature, curcumin undergo degradation into different products like vanillin, ferulic acid, and feruloyl methane.[18] The NMR studies by Payton et al. have shown that at pH range of 3–9, curcumin exists as keto-enol pH and Temperature Factor Affecting Curcumin Properties 173

tautomer in hydrophobic and hydrophilic solvents. Further agreements had given by Shen and Li by DFT calculations that calculated absorption and wavelength and oscillator strength of curcumin with experimental values, predominance of keto-enol tautomer. At 7.75 kcal^{-1}, ketoenol was found to be more stable than diketone. Attributed to planar structure of the enol form, stabilizes the resonance and twisted the ketone form.

(b)

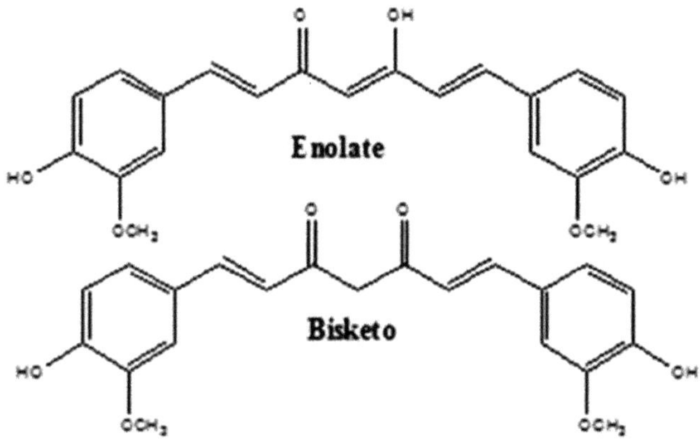

FIGURE 12.3 Chemical structure of curcumin.
Source: Reprinted with permission from Ref. [45]. © 2016 Korean Institute of Chemical Engineers, Seoul, Korea.

The growth of AgNPs with Cydonia oblonga seed is also effected by pH. By varying the pH values, various color intensity have been observed. Highest color intensity with monodispersive silver nanoparticles was recorded at pH 9. The SPR peaks show the rate of reduction increases with time (Fig. 12.3).

Curcumin conjugated gold nanoparticles: Synthesis of AuNPs in the presence of bio-molecules like soyabean, red cabbage, and bioreductants like tannic acid, Rutin have low molar extinction value than curcumin reducing potency, that is, 2. But in case of hyaluronic acid-curcumin, conjugated AuNPs give high yield, here hyaluronic acid acts as reducing agent, not curcumin. To enhance the curcumin reducing and stabilizing ability, increase in pH is very essential.

In human serum albumin (HSA, when reacting with curcumin, cancels the π conjugation, when the two vinyl guaiacol parts rotated curcumin to HSA.

By increase in pH of HAuCl$_4$ and further additions of curcumin results low SPR leads to poor AuNP formation, which, may be due to poor reactivity of acid at high alkalinity.

Figure 12.4 shows at pH 10.6, the SPR peak is broad even after 4 h of experiment, forming black colloids which indicates the inability of curcumin to complete reduction of Au^{3+} to Au^0. This is because at high pH, proton of water interacted with beta-diketone group of curcumin. Forming hydrogen bond with anions makes it inefficient for the reduction of Au^{3+}, which is further confirmed by TEM of NPs. It may be due to incomplete cleavage of larger intermediates to smaller particles since all the curcumin, molecules are not in Cur^{3-} form to reduce Au^{3+}. Pie et al, reported the same aggregation of unstable AuNPs to form nanowire-like structure.

Factor like pH ranges from 9.2 to 9.6 increases the wavelength from 428 nm to 438 nm which results in solubility of curcumin as well as deprotonation of enolic and phenolic group and exist as Cur^{3-}.[45]

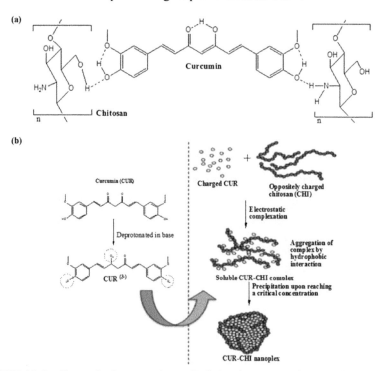

FIGURE 12.4 Curcumin deprotonation to Cur^{3-} forming aggregation complex.
Source: Reprinted with permission from Ref. [45]. © 2016 Korean Institute of Chemical Engineers, Seoul, Korea

pH and Temperature Factor Affecting Curcumin Properties

These two groups (phenolic and enolic groups) favors in reduction of Au^{3+} to Au^0. This is due to their pKa values which ranges from 7 to 10.6. Mild reducing agents like citrate capped Au seeds are used for the synthesis of nanorods. It was confirmed by adjusting the pH to its pKa value 9.3 to 9.6. Instead of nanorods, much smaller spherical particles were observed, further supported by blue shift in UV-Vis spectrum at 505 nm.[43]

Formation of citrate-stabilized AuNPs also depends upon pH, which should be above the pKa value of sodium citrate. pH value less than 9.2 does not support the formation of nanoparticles because of poor solubility of curcumin. Also curcumin gets unstable if pH increases to 11 and it leads to precipitation of cAuNPs (Fig. 12.5).[44]

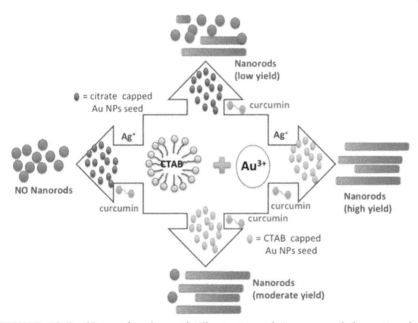

FIGURE 12.5 (See color insert.) Illustration of Au nanorod formation by using curcumin as secondary reducing agent through seed mediated method. *Source*: Reprinted with permission from Ref. [43]. © 2015 American Chemical Society.

12.2.2 pH EFFECT ON DRUG LOADING AND RELEASING WITH PDEA BLOCK

The pH also effects drug loading and releasing of drug. Rate of drug release was enhanced by pH value changed from 7.4 to 5.0. pH at 5.0, formation of hydrophilic segments through Protonation of curcumin loaded micelles that is, PDEA block, which resulted in faster release of curcumin.[20]

Reluctance of drug release in the cores medium at pH 7.4, because of no protonation of PDEA block which remained in compact form. But at pH 5.0, PDEA block swelled and ruptured leading to protonation of amine groups from PDEA blocks. Average size of curcumin loaded micelles was 182.2 nm measured by |DLS, which was larger than empty micelles.

12.2.3 *IN VITRO CURCUMIN RELEASE STUDY WITH INVITE MICELLES*

Fusion of hydrophobic drug is the most acceptable method into micellar body like Inulin-D-α-Tocop–herol Succinate (INVITE). It is precious for intravenous application in drug delivery systems. A hypothesis is there, INVITE micelle cores are formed by pi–pi interaction of aromatic portion of VITE rather than hydrophobic interaction between the aliphatic chains.[21–22] Curcumin with aromatic ring in the INVITE micelles does not form permanent stable system. So, without bursting effect encapsulated drug could be released in a controlled manner.[21]

The releasing behavior of the INVITE at pH 7.4 or 5.5, in a physiologic fluids or in a condition stimulating the endosome or the tumor environments had been studied in. Curves are showing about the release of free drug, curcumin, and the controlled release of curcumin in INVITE micelles system without bursting effect at both pH 7.4 or pH 5.5 in PBS solution in the presence of 0.5% polysorbate 80 up to 48 h.

A comparison was figured out that 42% from INVITE 3MC, 23% from INVITE 2MC, and 15% from INVITE 1MC after 48 h in PBS at pH had been released. At pH 5.5, there was 10% increase in value, that is, 53% for INVITE 3MC, 23% for INVITE 2MC, and 25% for INVITE 1MC.

This was due to "Bridge like structure" of VITE in INVITE 3M with hydrophobic portion of curcumin, the micelles resulted into formation of larger particles. Since, from Figure 12.1, TEM study shows there was no size difference between drug loaded and empty micelles. Further, it was noted that, release rate of drug was almost constant at no time fluctuations which confirmed the higher physical stability.[23,24] Similarly, a noticeable burst effect of curcumin has been reported by Abouzeid et al. and Sun

pH and Temperature Factor Affecting Curcumin Properties

et al.[25] Hence, rate of release at pH 7.4 or 5.5 may be useful for tumor delivery in acidic environment (Figs. 12.6 and 12.7).

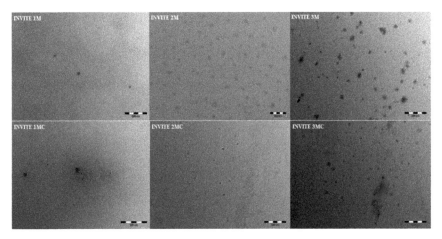

FIGURE 12.6 TEM images of empty INVITE M's and curcumin loaded INVITE MC's of equal size of 200 nm.
Source: Reprinted with permission from Ref. [21]. © 2015 American Chemical Society.

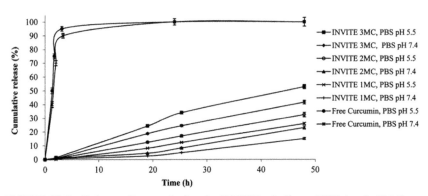

FIGURE 12.7 Release of curcumin drug by INVITE micelles at PH 7.4 and pH 5.5.
Source: Reprinted with permission from Ref. [21]. © 2015 American Chemical Society.

12.2.4 EFFECT OF TEMPERATURE

Incubation at 37°C with 10% serum, human blood, PBS solution, culture medium was done. Stability of curcumin in serum or human blood was observed. Under serum free condition, curcumin undergoes degradation into vanillin product. So, care should be taken for storage of curcumin. −20°C is the storage temperature for curcumin. But there is no evidence for instability of curcumin at higher temperature. In another study by Gopinath et al., micro shrinkage of pure collagen fiber and curcumin containing collagen fiber 55°C was done and it was observed that collagen fiber got shrinked which further increased up to 78°C. That means curcumin polyphenol increases the stability of collagen, which could be used to cure dermal wound healing. Previous studies revealed that heating causes the degradation, structural changes, and loss of biological activities in curcumin. Unmodified HEWL (Hen egg white lysosome) and curcumin-treated HEWL incubated at 70°C show similar results. Also, it was observed at 55°, not all curcumin was degraded. It was clearly shown by 50% decrease in fluorescence result. But at 70°C temperature, higher amount of amyloid fibrils was observed in modified and unmodified HEWL. So, At 70°C, curcumin degraded completely into further products within 24 h. Temperature range between 10°C and 55°C suitable for curcumin thermal stability.

Thermograms were recorded for plain poly (AM-co-AMPS) hydrogel.[26] Silver ion loaded hydrogel, silver nanocomposite hydrogel, and curcumin loaded silver nanocomposite hydrogels are shown in Figure 12.4a and c, respectively. From the Figure 12.4, it is observed that all the three samples start degrading at 150°C. It was also observed that, at 700°C, 100% weight loss is noticed for PAM+A4 hydrogel but in PAM-A4+Ag$^+$ hydrogel 18% of residue was found. But in the case of PAM-A4+Ag^0 hydrogel the more residue is (25%) noticed, because of formation of silver nanoparticles in the hydrogel network. When we observe the thermograms of curcumin loaded hydrogel samples they also show similar trend as that of empty hydrogels. But curcumin loaded silver nanocomposite hydrogels show higher thermal stability.

It has been found that via Differential scanning calorimetry that PAM-A4, PAM-A4 Ag+, and PAM-A4+Ag0 hydrogels attains the temperature of 76.5°C, 75.2°C, and 98.4°C. Formation silver nanoparticles increases the melting temperature of hydrogel samples. Similarly, the curcumin loaded hydrogel samples shows almost similar behavior. However, because of the presence of curcumin, the curcumin loaded hydrogels have shown an additional peak between 175°C and 201°C due to melting temperature of curcumin.

12.3 BIOAPPLICABILITY

12.3.1 CURCUMIN LOADING AND RELEASE STUDIES WITH Ag NANOCOMPOSITE HYDROGELS[26]

Figure 12.8 resulted in the encapsulation efficiency of loading and releasing drug curcumin from which it was observed AgNPs loaded hydrogels had higher efficiency than simple hydrogels like plain poly (AM-co-AMPs) and Ag+ loaded hydrogel.

Reason behind least loading efficiency of Ag+ ions loaded hydrogels was AMP chains which were interacted by Ag+ ions and therefore inhibit the loading capacity of drug into the hydrogels.

For drug release, experiments were conducted at pH 7.4 of buffer solution at 37°C, on plain poly(AM-co-AMPs) hydrogel, silver nanocomposite hydrogels and curcumin loaded silver nanocomposite hydrogel. It was observed that with interval of time curcumin loaded silver nanocomposite had slower release of drug than plain poly(AM-co-AMPs) hydrogel and AgNPs loaded hydrogels because of presence of number of curcumin molecules adsorbed on silver NPs in addition to entrapment in the hydrogels.

Further antibacterial activity was checked on plain poly(AM-co-AMPs) hydrogel, HSNC, and curcumin loaded HSNC's against *E. coli* bacteria. It was observed that plain hydrogel did not show any effect on bacterial growth. Both HSNC's and curcumin loaded HSNC's suppressed the growth of bacteria.

Order of antibacterial activity is as follows:

Curcumin loaded HSNC > HSNC's > plain poly(AM-co-AMPs) hydrogel

This was due to curcumin and silver NPs that the growth of bacteria was suppressed and they also acted as superior antibacterial materials and wounds' cure.

12.3.2 IN VITRO CURCUMIN RELEASE AND LOADING WITH % MMT[27]

The kinetic parameters had been studied regarding release of curcumin from nanoformulations. By applying regression co-efficient analysis

($R^2 > 0.95$), power law reported the mechanism regarding curcumin release from nanoformulation from stimulated intestinal fluid with a faster release of curcumin at initial stage than later stage. Besides this, percent composition of MMT encapsulation also affected the release of curcumin drug, means, 15% MMT formulation revealed faster drug release than 30% MMT and without MMT nanoformulation encapsulation.

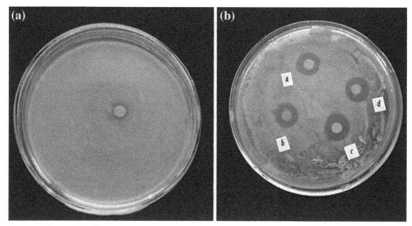

FIGURE 12.8 Antibacterial activity with curcumin-loaded HSNC.
Source: Reprinted with permission from Ref. [26]. © 2015 American Chemical Society.

The drug loading percentage and encapsulation efficiency percentage of curcumin nanoparticles appeared to be high for 15% MMT with formulation than formulation containing 0% and 30% MMT.

For drug release, size of nanoparticle is the important factor that plays main role in drug delivery system. Curcumin nanoformulation with MMT had larger sizes than zero MMT nanoformulation. As nanoparticles size is directly related to diffusion rate of organic solvent to outer aqueous medium. Formulation containing MMT had high viscosity that resulted to larger size nanoparticles. This high encapsulation efficiency lead to availability of drug at target site and increased residence time of drug. Release of drug with MMT formulation due to presence of nanoclay that inhibited the diffusion of curcumin and slowed down their aggregation which further lead to their dissolution. 15% of MMT formulation had faster release of drug than 30% because higher concentration of MMT decrease the penetration of water into the PCL nanocomposite and inhibits the erosion of the polymer matrix.

Cur/MPEGPCL[28] micelles have stronger anticancer effect than free curcumin. This means that if we encapsulate curcumin with MPEG-PCL micelles, it will enhance the anticancer activity of curcumin in vivo. Curcumin has cytotoxic effects on cancer cells, which can kill cancer cells in vivo directly. In addition to anticancer property, curcumin inhibits angiogenesis (formation of new blood vessels which is necessary for tumor growth) which also results in inhibition of tumor.[28–36,37,38] MPEG-PCL micelles have potential application in drug delivery systems and are already used to deliver some drugs. MPEG-PCL was simple and easy to scale up. MPEG-PCL micelles were monodisperse with a mean particle size of 27.3–1.3 nm and had encapsulation efficiency of 99.16–1.02% and a drug loading of 12.95–0.15%. Some experiments were done for many days that is, vasculation of alginate implants on mice, in which Alginate beads having colon carcinoma cells were injected subcutaneously into the mice on day 0. Then, these mice were managed with normal saline intravenously, MPEG-PCL micelles, free curcumin, or Cur/MPEG-PCL micelles were given once a day for 5 days. On day 12, beads were surgically removed, FITC-dextran was also used on mice and it was observed that there was reduction of vascularization in mice which was treated with Cur/MPEG-PCL micelles.[39]

Nanocurcumin (NC), unformulated curcumin, was experimented once a week for 3 months to 9-month-old Tg2576[40] mice and its changes in memory were measured by radial arm maze (RAM) and contextual fear conditioning tests. At the end of the study, amyloid plaque burden in the mouse brains was compared among the three treatment groups this may be biphasic dose–response of curcumin on proteasome activity. Low concentrations of curcumin enhance proteasome activity, but higher concentrations minimize proteasome activity. Curcumin also blocks Aß aggregation by direct binding to Aß, and this effect is monophasic, with greater effect at greater concentrations.[41,42] Perhaps proteasome inhibition increases protein aggregation and Aß plaques, while Aß binding decreases formation of oligomers that affect neuronal function. Low concentrations of curcumin may thus reduce plaques and protect neuronal function, but the higher concentrations achieved by NC may produce less plaque reduction but better protection of neuronal function.

12.4 CONCLUSION

In summary, simple and progressive encapsulation method is adopted to enhance the bioavailability of curcumin clinically. Solid dispersion methods are used to increase the solubility of curcumin in aqueous medium. Various encapsulation methods have been introduced nowadays, like encapsulation with INVITE micelles, AgNP hydrogels network, functionalized silica nanoparticles, MMT formulation, and so on. Besides this, factor like pH and temperature and curcumin concentration and % formulation also affect the solubility, stability, and efficiency of release of drug in vitro system. Even studies showed that green chemistry used for formation of nanoparticles like AgNPs with Cydonia oblonga seeds extract which acts as both capping as well as stabilizing agent. Nano encapsulation with controlled environment loading and release of curcumin drug can be maintained. As a result, this nano encapsulation can be directly used for antibacterial and cure of wound. Curcumin loaded materials enhances the wound healing, prolongs drug transit time, improves solubility, and sustain release of curcumin to targeted area, that is, target drug delivery.

KEYWORDS

- **bioavailability**
- **curcumin**
- **encapsulation**
- **hydrophobic**
- **nanocarrier**
- **pH**

REFERENCES

1. Bharath, L. D.; Anushree, K.; Agarwal, M. S.; Shishodia, S. Curcumin Derived from Turmeric Curcuma longa. *Phytochem. Cancer Chemopreven.* 349–387.
2. Negi, P. S.; Jayaprakash, G. K.; Jagan, M. R. L.; Sakariah, K. K. Antibacterial Activity of Turmeric Oil; A Byproduct From Curcumin Manufacturer. *J. Agric. Food Chem.* **1999**, *47*, 4297–300.

3. Apisariyakul, A.; Vanittanakomm, N.; Buddhasukh, D. Antifungal Activity of Turmeric Oil Extracted from Curcuma Longa (Zingiberaceae). *J. Ethnopharmacol.* **1995**, *49*, 163–169.
4. Ghatak, N.; Basu, N.; Sodium Curcuminate as an Effective Anti-inflammatory Agent. *Ind. J. Exp. Biol.* **1972**, *10*, 235–236.
5. Itthipanichpong, C.; Ruangrungsi, N.; Kemsri, W.; Sawasdipanich, A. Antispasmodic Effects of Curcuminioids on Isolated Guinea Pig Ileum and Rat Uterus. *J. Med. Assoc. Thai.* **2003**, *86*, 299–309.
6. Ruby, A. J.; Kuttan, G.; Dinesh, B. K.; Rajasekharan, K. N.; Kuttan, R. Antitumor and Antioxidant Activities of Natural Curcuminoids. *Cancer Lett.* **1995**, *94*, 79–83.
7. Dhar, M. L.; Dhar, M. M.; Dhawan, B. N.; Malhotra, B. N.; Ray, C. Screening of Indian Plants for Biological Activity. *Ind. J. Exp. Biol.* **1968**, *6*, 232–247.
8. Mazumdar, A.; Raghavan, K.; Weinstein, J.; Kohn, K. W.; Pommer, Y. Inhibition of Human Immunodeficiency Virus Type-1 Integrase by Curcumin. *Biochem. Pharmacol.* **1995**, *49*, 1165–1170.
9. Halim, E. M.; Ali, H.; Hypoglycemic, Hypolipidemic and Antioxidant Properties of Combination of Curcumin from Curcuma Longa Lin. and Partially Purified Product from Abroma Augusta Lin. in Streptozotocin Induced Diabetes. *Ind. J. Clin. Biochem.* **2002**, *17* (12), 33–43.
10. Garg, S. K. Effect of Curcuma Longa (Rhizomes) on Fertility in Experimental Animals. *Planta Med.* **1974**, *26*, 225–227.
11. Ling, G.; Wei, Y.; Ding, X. Transcriptional Regulation of Human CYP2A13 Expression in the Respiratory Tract by CCAAT/Enhancer Binding Protein and Epigenetic Modulation. *Mol. Pharmacol.* **2006**, *71* (3), 807–816.
12. Okonogi, S.; Oguchi, T.; Yonemochi, E.; Puttipipatkhachorn, S.; Yamamoto, K. Physicochemical Properties of Ursodeoxycholic Acid Dispersed in Controlled Pore Glass. *J. Coll. Interface Sci.* **1999**, *216*, 276–84.
13. Kapoor, L. D. *Handbook of Ayurvedic Medicinal Plants*; CRC Press: Boca Raton, Florida, 1990; pp 185.
14. Modasiya, M. K.; Patel, V. M. Studies of Solubility of Curcumin. *Int. J. Pharm. Sci.* **2012**, *3*, 1490–1497.
15. Kiuchi, F.; Goto, Y.; Sugimoto, N.; Akaon, N.; Kondo, K.; Tsuda, Y. Nemotodial Activity of Turmeric. *Chem. Pharm. Bull. Tokyo* **1993**, *41* (A), 1640–1643.
16. Ravindranath, V.; Chandrasekhara, N. In Vitro Studies on the Intestinal Absorption of Curcumin in Rats. *Toxicology* **1981**, *20*, 251–257.
17. Gou, M. L.; Zheng, X. L.; Men, K.; Zhang, J.; Wang, B. L.; L. L. et al. Self-assembled Hydrophobic Honokiol Loaded MPEG-PCL Diblock Copolymer Micelles. *Pharm. Res.* **2009**, *26*, 2164–2173.
18. Wang, Y. J.; Pan, M. H.; Chang, A. L.; Hsieh, C. Y.; Lin, J. K. Stability of Curcumin in Buffer Solutions and Characterization of its Degradation Products. *J. Pharm. Biomed. Anal.* **1997**, *15*, 1867–1876.
19. Newa, M.; Bhandari, K.; Jong, O. K.; Seob, J.; Jung, Ae.; Bong, K.; Kyu, Y.; Jong, Soo W.; Han Gon, C.; Chul, Soon, Y. Enhancement of Solubility, Dissolution and Bioavailability of Ibuprofen in Solid Dispersion Systems. *Chem. Pharm. Bull.* **2008**, *56* (4), 569–574.

20. Zhai, S.; Ma, Y.; Chen, Y.; Li, D.; Cao, J.; Liu, Y.; Cai, M.; Xie, X.; Chen, Y.; Luo, X. Synthesis of an Amphiphilic Block Copolymer Containing Zwitterionic Sulfobetaine as a Novel Ph-Sensitive Drug Carrier. *Poly. Chem.* **2014,** *5,* 1285–1297.

21. Tripodo, G.; Pasut, G.; Trapani, A.; Mero, A.; Lasorsa, M. F.; Chlapanidas, T.; Trapani, G.; Mandracchia, D. Inulin-D-A-Tocopherol Succinate (INVITE) Nanomicelles as a Platform for Effective Intravenous Administration of Curcumin. *Biomacromolecules* **2015,** *16,* 550–557.

22. Tripodo, G.; Mandracchia, D.; Dorati, R.; Latrofa, A.; Genta, I.; Conti, B. Nanostructured Polymeric Functional Micelles for Drug Delivery Applications. *Macromol. Symp.* **2013,** *334* (1), 17–23.

23. Catenacci, L.; Mandracchia, D.; Sorrenti, M.; Colombo, L.; Serra, M.; Tripodo, G. In-solution Structural Considerations by 1H NMR and Solid-state Thermal Properties of Inulin-D-A-Tocopherol Succinate (INVITE) Micelles As Drug Delivery Systems for Hydrophobic Drugs. *Macromol. Chem. Phys.* **2014,** *215,* 2084–2096.

24. Mandracchia, D.; Tripodo, G.; Latrofa, A.; Dorati, R. Amphiphilic Inulin-D-A-Tocopherol Succinate (INVITE) Bioconjugates for Biomedical Applications. *Carbohydr. Polym.* **2014,** *103,* 46–54.

25. Abouzeid, A. H.; Patel, N. R.; Torchilin, V. P. Polyethylene Glycol Phosphatidylethanolamine (PEG-PE)/Vitamin E Micelles for Co-delivery of Paclitaxel and Curcumin to Overcome Multi-drug Resistance in Ovarian Cancer. *Int. J. Pharm.* **2014,** *464* (1–2), 178–184.

26. Ravindra, S.; Mulaba-Bafubiandi, F. A.; Rajinikanth, V.; Varaprasad, K.; Reddy, N. N.; Mohana Raju, K. Development and Characterization of Curcumin Loaded Silver Nanoparticle Hydrogels for Antibacterial and Drug Delivery Applications. *J. Inorg. Organomet. Polym.* **2012,** *22,* 1254–1262.

27. Bakre, G. L.; Sarvaiya, I. J.; Agarwal, K. Y.; Synthesis, Characterization, and Study of Drug Release Properties of Curcumin from Polycaprolactone/Organomodified Montmorillonite Nanocomposite. *J. Pharma. Innov.* **2016,** *11* (4), 300–307.

28. Gou, M.; Men, K.; Shi, H.; Xiang, M.; Zhang, J.; Song, J.; Long, J.; Wan, Y.; Luo, F.; Zhao, X.; Qian, Z. Curcumin-loaded Biodegradable Polymeric Micelles for Colon Cancer Therapy In Vitro and In Vivo. *Nanoscale* **2011,** *3,* 1558.

29. Lin, J. K.; Lin-Shiau, S. Y. Mechanisms of Cancer Chemoprevention by Curcumin. *Proc. Natl. Sci. Counc. ROC (B)* **2001,** *25,* 59–66.

30. Rao, C. V.; Rivenson, A.; Simi, B.; Reddy, B. S. Chemoprevention of Colon Carcinogenesis by Dietary Curcumin, a Naturally Occurring Plant Phenolic Compound. *Cancer Res.* **1995,** *55,* 259–66.

31. Kawamori, T.; Lubet, R.; Steele, V. E.; Kelloff, G. J.; Kaskey, R. B.; Rao, C. V. et al. Chemopreventive Effect of Curcumin, a Naturally Occurring Anti-inflammatory Agent, During the Promotion/Progression Stages of Colon Cancer. *Cancer Res.* **1999,** *59,* 597–601.

32. Dorai, T.; Cao, Y. C.; Dorai, B.; Buttyan, R.; Katz, A. E. Therapeutic Potential of Curcumin in Human Prostate Cancer. III. Curcumin Inhibits Proliferation, Induces Apoptosis, and Inhibits Angiogenesis of lnCap Prostate Cancer Cells in vivo. *Prostate* **2001,** *47,* 293–303.

pH and Temperature Factor Affecting Curcumin Properties   185

33. Mukhopadhyay, A.; Bueso-Ramos, C.; Chatterjee, D.; Pantazis, P.; Aggarwal, B. B. Curcumin Downregulates Cell Survival Mechanisms in Human Prostate Cancer Cell Lines. *Oncogene* **2001,** *20,* 7597–609.
34. Choudhuri, T.; Pal, S.; Agwarwal, M. L.; Das, T.; Sa, G. Curcumin Induces Apoptosis in Human Breast Cancer Cells Through P53dependent Bax Induction. *FEBS Lett.* **2002,** *512,* 334–40.
35. Li, L.; Braiteh, F. S.; Kurzrock, R. Liposome-encapsulated Curcumin in vitro and in Vivo Effects on Proliferation, Apoptosis, Signaling, and Angiogenesis. *Cancer* **2005,** *104,* 1322–31.
36. Bharti, A. C.; Donato, N.; Singh, S.; Aggarwal, B. B. Curcumin (Diferuloylmethane) Down-regulates the Constitutive Activation of Nuclear Factor-Kb and IKba Kinase in Human Multiple Myeloma Cells, Leading to Suppression of Proliferation and Induction of Apoptosis. *Blood* **2003,** *101,* 1053–62.
37. Tian, B.; Wang, Z.; Zhao, Y.; Wang, D.; Li, Y.; Ma, L. et al. Effects of Curcumin on Bladder Cancer Cells and Development of Urothelial Tumors in a Rat Bladder Carcinogenesis Model. *Cancer Lett.* **2008,** *264,* 299–308.
38. Arbiser, J. L. Klauber, N.; Rohan, R.; Van Leeuwen, R.; Huang, M. T.; Fisher, C. et al. Curcumin is an in vivo Inhibitor of Angiogenesis. *Mol. Med.* **1998,** *4,* 376–83.
39. Goel, A.; Kunnumakkara, A. B.; Aggarwal, B. B. Curcumin as "Curecumin": From Kitchen to Clinic. *Biochem. Pharmacol.* **2008,** *75,* 787–809.
40. Kwok Kin Cheng, K. K.; Yeung, F. C.; Ho, W. S.; Chow, F. S.; Chow, L. H. A.; Baum, L. Highly Stabilized Curcumin Nanoparticles Tested in an in vitro Blood–Brain Barrier Model and in Alzheimer's Disease Tg2576 Mice. *AAPS J.* **2013,** *15,* 2.
41. Yang, F.; Lim, G. P.; Begum, A. N.; Ubeda, O. J.; Simmons, M. R.; Ambegaokar, S. S. et al. Curcumin Inhibits Formation of Amyloid ß Oligomers and Fibrils, Binds Plaques, and Reduces Amyloid in vivo. *J. Biol. Chem.* **2005,** *280,* 5892–901.
42. Begum, A. N.; Jones, M. R.; Lim, G. P.; Morihara, T.; Kim, P.; Heath, D. D. et al. Curcumin Structure-function, Bioavailability, and Efficacy in Models of Neuroinflammation and Alzheimer's Disease. *J. Pharmacol. Exp. Ther.* **2008,** *326,* 196–208.
43. Moussawi, N. R.; Patra, D. Synthesis of Au Nanorods Through Prereduction with Curcumin: Preferential Enhancement of Au Nanorod Formation Prepared from CTAB-Capped over Citrate-Capped Au Seeds. *J. Phys. Chem. C* **2015,** *119,* 19458–19468.
44. Hubert, F.; Testard, F.; Spalla, O. Cetyltrimethylammonium Bromide Silver Bromide Complex as the Capping Agent of Gold Nanorods. *Langmuir* **2008,** *24,* 9219–9222.
45. Mahmood. K. et al. Enhancement of Bioactivity and Bioavailability of Curcumin with Chitosan Based Materials. *Korean J. Chem. Eng.* **2016,** *33* (12), 3316–3329.

CHAPTER 13

Integrated Water Resource Management and Nanotechnology Applications in Water Purification: A Critical Overview

SUKANCHAN PALIT*

43, Judges Bagan, Post Office Haridevpur, Kolkata 700082, India

Corresponding author. E-mail: sukanchan68@gmail.com

ABSTRACT

The world of environmental engineering science and chemical process engineering are gearing forward toward newer vision and a newer scientific regeneration. Water purification, drinking water treatment, and industrial wastewater treatment are the important issues facing the human civilization today. Today water treatment stands in the midst of deep scientific ingenuity and vast scientific profundity. In the similar vision, integrated water resource management is the utmost need of the hour as human civilization and human scientific endeavor moves forward. Today nanotechnology and water purification are challenging the vast global scientific firmament. In this treatise, the author deeply elucidates the scientific success and the scientific provenance behind the application of nanotechnology in water purification and environmental engineering. Mankind's immense scientific prowess, girth and determination, man's immense scientific provenance, and the futuristic vision of environmental engineering, chemical engineering, and nanotechnology will definitely lead an effective way in the true realization of engineering science and technology. The author deeply delineates the different innovations of environmental engineering and water purification techniques globally.

The true emancipation of science and engineering will be emboldened as the author moves from one visionary frontier toward another in the field integrated water resource management. Water resource management and water purification are the imminent needs of human scientific progress. Arsenic and heavy metal groundwater contamination are veritably challenging the deep scientific fabric globally. The main pillars of this treatise are the vast nanotechnology applications in water purification and the vast and versatile research and development forays in the field of integrated water resource management. Integrated water resource management and the recent advances in water purification are the other pillars of this treatise. The author also depicts profoundly the vast scientific vision behind environmental sustainability and its immense scientific relevance.

13.1 INTRODUCTION

Today, science and human civilization are moving at a rapid pace globally. Technology and engineering science in the similar vein are in the path of newer scientific regeneration. Environmental engineering science, water purification, and nanotechnology are the branches of scientific endeavor which needs to be re-envisioned and reorganized with the passage of scientific history and time. Depletion of fossil fuel resources, loss of ecological biodiversity, and frequent environmental disasters are veritably challenging the scientific fabric globally. The author, in this treatise, deeply elucidates the success of science, the vast scientific profundity and the futuristic vision of the field of water purification, drinking water treatment, and industrial wastewater treatment. Today, energy and environmental sustainability are the utmost needs of humanity and human scientific progress. Sustainable development whether it is social, economic, energy, or environment is the imminent need of scientific progress globally. Global water crisis and global climate change are the vexing issues of human civilization. Technology, engineering, and science globally are in a state of deep distress and immense comprehension. In this chapter, the author pointedly focuses on the vision, success, targets, and innovations in the field of environmental engineering and nanotechnology. The other visionary areas of scientific introspection are integrated water resource management.

Integrated Water Resource Management and Nanotechnology Applications   189

13.2   THE AIM AND OBJECTIVE OF THIS STUDY

The main aim and objective of this study is to investigate nanotechnology applications in environmental engineering science and the vast world of applications of integrated water resource management in human scientific progress. Today, human factor engineering and systems science are connected to water purification and water resource management. The author pointedly focuses on these aspects of research pursuit with a clear vision toward furtherance of science and engineering. Success of science globally depends upon sustainability, effective mitigation of global climate change, and the application of nanoscience and nanotechnology. The author also delineates the recent and significant advances of application of nanotechnology in water purification, industrial wastewater treatment, and water resource management. Global scenario in the field of water purification, drinking water treatment, and industrial wastewater treatment is extremely dismal. Integrated water resource management is one of the feasible solutions to global water issues. Thus, the world of science and engineering will open up new doors of innovation and scientific instinct in the field of water resource engineering and environmental engineering in decades to come. Nanotechnology and nano-engineering are veritably opening up newer scientific intricacies and scientific and engineering hurdles. The author, in this treatise, deeply pronounces the success of nanotechnology and nano-engineering in confronting global water issues, global climate change, and the frequent environmental disasters.

13.3   THE SCIENTIFIC DOCTRINE OF INTEGRATED WATER RESOURCE MANAGEMENT

Today, integrated water resource management is a veritable pillar of environmental engineering and water resource engineering. The scientific doctrine of integrated water resource management is linked with every branch of environmental engineering. This treatise is a vast eye-opener toward the immense scientific potential of water resource management and water purification. Environmental engineering science is in the midst of scientific vision and deep scientific forbearance. Water is an integral part of human society and is a major component of the scientific progress of human civilization. Water, electricity, energy, food, education,

190   Green Chemistry and Biodiversity: Principles, Techniques, and Correlations

and shelter are the imminent needs of human civilization. In the similar vein, energy and environmental sustainability stand as an integral part toward the progress and advancement of human civilization. Here comes the need and the importance of integrated water resource management, water resource engineering, and the holistic domain of environmental engineering science.

13.4 THE VISION OF MEMBRANE SCIENCE AND OTHER NOVEL SEPARATION PROCESSES

Water science and technology and its vision are the immediate need of the hour. Membrane science and novel separation processes are revolutionizing the scientific landscape and changing the face of human scientific research pursuit. Reverse osmosis, nanofiltration, and forward osmosis are changing the scientific fabric of water purification and separation science. Today, the world of science is highly concerned with frequent environmental disasters, loss of biodiversity, and water pollution control. Here comes the immediate need of membrane science and novel separation techniques. Forward osmosis is solving many scientific barriers in water pollution control and industrial wastewater treatment. Scientific community is at its foot in solving major industrial pollution control issues and the scientific progress of human race needs to be re-envisioned as engineering science and technology breaks barriers.

13.5 ENVIRONMENTAL SUSTAINABILITY AND THE VISION FOR THE FUTURE

Water is mankind's most important resource—water is the life of human civilization. Humans need water for drinking, for washing, for irrigation of croplands, for the production of commodities, and cooling of power plants. Environmental sustainability and renewable energy technology stand as major pillars toward the scientific success of human civilization. The vision for the future in the field of environmental engineering science needs to be re-organized as mankind surges forward toward newer scientific might and scientific determination. Today, nanotechnology is a marvel of scientific research pursuit. The author, in this paper, deeply pronounces

Integrated Water Resource Management and Nanotechnology Applications   191

the integration of nanotechnology and environmental engineering in the clear vision toward furtherance of science and technology globally.

13.6   ARSENIC AND HEAVY METAL GROUNDWATER REMEDIATION AND THE FUTURE OF WATER QUALITY

Arsenic and heavy metal water poisoning are challenging the vast scientific fabric of developing and developed nations around the world. Water standards and water quality are the imminent needs of scientific ardor and vision globally. Here also comes the vast importance of water resource management and water resource engineering. A multi-disciplinary approach of science and engineering is the need of the hour. Water quality and its scientific inquiry are of utmost importance in modern science today. In South Asia, particularly Bangladesh and India, heavy metal groundwater poisoning is a major environmental disaster. Human civilization's immense scientific perseverance, futuristic vision, and deep scientific retrospection will veritably unlock the intricacies and barriers of groundwater remediation research. Today, human scientific challenges in the field of environmental remediation are immense and thought-provoking as environmental sustainability faces a deepening crisis. Sustainability science is highly retrogressive in modern civilization as environmental remediation challenges the vast scientific firmament of scientific might and vision. Geologists, environmental engineers, chemical engineers, and environmental economists are gearing toward a newer era in environmental protection and drinking water treatment. A lasting impression in interdisciplinary science and engineering is needed as global climate change and global warming moves forward toward a disastrous era. Bangladesh and many states in India are in the throes of a major health disaster and an unmitigated environmental catastrophe. Groundwater remediation technology and water science and engineering need to be revamped and re-organized as nations around the world grapples with the burning issue of provision of pure drinking water and the concerns of environmental sustainability. In this entire treatise, the author pointedly focuses on the success of science and engineering in mitigation of global water issues and the success and emancipation of water engineering. The challenge and the vision of water technology are massive as well as far-reaching. Innovations, technological, and scientific validation are the pillars of scientific

192 Green Chemistry and Biodiversity: Principles, Techniques, and Correlations

research pursuit globally. In the similar vein, water resource engineering, environmental engineering science, and sustainable development will lead a long and effective way in the true unfolding of scientific truth of environmental protection and groundwater decontamination.

13.7 NANOTECHNOLOGY AND WATER PURIFICATION

Nanotechnology and water engineering are linked to each other by an umbilical cord. Systems engineering and human factors engineering are linked to diverse branches of science and engineering. Water purification and industrial wastewater treatment techniques are the boons of human civilization and environmental sustainability. In the similar vein, human factor engineering, systems, and industrial engineering are the veritable forerunners toward newer scientific vision and scientific determination. Nanofiltration and membrane separation tools are the needs of human society. Water purification, wastewater treatment, and water resource engineering are in the midst of deep scientific and environmental engineering crisis. The author in this chapter deeply pronounces the needs of integration of nanotechnology with water purification.

13.8 NANOFILTRATION, ULTRAFILTRATION, AND MICROFILTRATION

The science of nanofiltration, ultrafiltration, and microfiltration are on the path of newer vision and newer scientific struggle and deep perseverance. Membrane separation processes and other traditional and non-traditional environmental engineering tools are veritably challenging the vast scientific fabric of might, struggle, and vision. Water treatment and water purification are the utmost need of humanity and the global civil society. This chapter is a veritable eye-opener toward the present day necessities of human civilization which includes water, energy, sustainability, food, shelter, and education. The vision and the challenges of implementation of environmental and energy sustainability are immense and groundbreaking. Here comes the importance of novel separation processes and non-traditional environmental engineering techniques. Nanofiltration, ultrafiltration, and microfiltration are the veritable challenges of science

Integrated Water Resource Management and Nanotechnology Applications   193

and technology. Apart from water purification, nanofiltration, ultrafiltration, and microfiltration can be applied in food industry and dairy technology. The vast scientific vision, the mankind's immense scientific grit and determination, and the needs of human society will all lead a long and visionary way in the true realization of environmental engineering and chemical process engineering.

13.9 INTEGRATED WATER RESOURCE MANAGEMENT, HUMAN FACTOR ENGINEERING, AND NANOTECHNOLOGY

Human factor engineering and systems engineering are veritably linked with diverse branches of engineering science. In the similar vein, integrated water resource management and nanotechnology are today integrated with human factor engineering. The challenges and the vision of human factor engineering applications in water resource engineering and environmental remediation are vast and versatile. In this entire treatise, the author profoundly depicts the success of engineering science and technology in battling environmental engineering problems. Human factor engineering and systems engineering should be in today's global scenario, integrated with environmental engineering and other diverse areas. Science and engineering of nanotechnology should be emboldened as water crisis enhances. In this treatise, the author deeply depicts the needs of system science and nanotechnology in water purification.

13.10 RECENT SCIENTIFIC ADVANCES IN THE FIELD OF WATER PURIFICATION

Today, water purification and industrial wastewater treatment are the needs of human civilization and human scientific progress. Groundwater heavy metal contamination is a bane of human civilization. In this section, the author deeply elucidates on the success and vision of science and engineering in the field of water purification. Provision of pure drinking water and environmental sustainability are in a crisis as human civilization moves forward. This vital need of human civilization will reorganize the scientific thoughts and scientific inquiry of environmental protection and remediation.

194   Green Chemistry and Biodiversity: Principles, Techniques, and Correlations

Kanchi[1] deeply discussed with scientific farsightedness nanotechnology for water treatment. Nano-science and nanotechnology are the revolutionary areas of science and engineering. The term nanotechnology can be defined as a range of technologies performed on a nanometer scale with a widespread application areas.[1] The author, in this paper, deeply discusses significance of nanotechnology in wastewater treatment. Instrument-based wastewater treatment analysis with nanomaterials is the veritable pivot of scientific understanding and the vision of this paper. Nanomaterials and engineered nanomaterials applications in water purification are deeply investigated.[1] The deep scientific vision and the vast scientific challenges in the field of nano-adsorbents application are dealt lucidly.[1]

Sengupta et al.[2] deeply elucidates, with scientific conscience, a simple chemical-free arsenic removal method for community water supply as a case study from West Bengal, India. This report describes a simple chemical-free method that was successfully used by a team of European and Indian scientists to remove arsenic (As) from groundwater in a village in West Bengal, India.[2] The study was conducted in Kasimpore, a village in North 24 Parganas district, approximately 25 km from Kolkata. The total area of the village is 5.0 km and the annual income of the villagers is US$ 350/annum. The main source of water for the village is shallow wells and tube wells.[2] Kasimpore was chosen as the model village in the study since preliminary study revealed that 70% of the tube wells in this village had arsenic concentrations above 50.0 mg/L as compared to the WHO guideline value of 10 mg/L. Arsenic groundwater contamination in Bangladesh and the state of West Bengal in India is the world's most alarming and greatest environmental catastrophe.[2] According to one estimate, nearly 100 million rural people are affected by Arsenic exposure in food and drinking chain in Asia. Thus, there is the need of a sound environmental engineering technique.[2] Subterranean groundwater treatment is based on the principle of oxidation and filtration processes of conventional surface treatment plants for removal of Iron and Manganese from water.[2] Technological drive, the deep scientific inquiry, and sagacity are the necessities of scientific and technological revolution and deep re-envisioning today. In this case, underground aquifer is used as a natural biochemical reactor and adsorber, that removes As (arsenic) along with Fe (iron) and Mn (manganese) at an elevated redox value of groundwater when dissolved oxygen concentration is raised above 4 mg/L.[2] Water is pumped from the underground aquifer using a submersible pump.[2] About 15–20% of this

Integrated Water Resource Management and Nanotechnology Applications   195

water is returned to the aquifer at the same depth under gravity while the remaining water is supplied as an arsenic free to the customer of the water supply. Employing this technology, the team has installed six plants in West Bengal, India for supplying potable water to rural communities.[2] A new technology thus opened up newer scientific vision and scientific envisioning in groundwater remediation and environmental protection.

Hashim et al. (2009)[3] deeply elucidated with cogent insight remediation technologies for heavy metal contaminated groundwater. The contamination of groundwater by heavy metal, originating from natural soil resources or from anthropogenic sources is a matter of immense scientific concern for public health and hygiene. In this paper, 35 approaches for groundwater treatment have been reviewed and classified under three large categories which are chemical, biochemical/biological/biosorption, and physico-chemical treatment processes.[3] Selection of a suitable technology for groundwater contamination remediation at a particular site is one of the most challenging areas of water science and water engineering. Keeping the environmental sustainability issues in mind, the technologies encompassing natural chemistry, bioremediation, and biosorption are highly recommended to be adopted in particular environmental engineering problem.[3] In many places, two or more tools can work synergistically for better results.[3] Process such as chelate extraction and chemical soil washings are highly advisable for recovery of valuable metals in contaminated industrial sites. In this review, the authors deeply discussed sources, chemical property and speciation of heavy metals in groundwater. Technological revamping and vast re-envisioning are the pillars of this treatise. The technologies discussed are (1) chemical treatment technologies, (2) reduction by dithionite, (3) reduction by iron-based technologies, (4) soil flushing, (5) in-situ soil flushing, (6) in-situ chelate flushing, and (7) other biological, biochemical, and biosorptive treatment technologies.[3] In this paper, biosorption of heavy metals is also discussed in minute details.[3]

Hassan (2018)[4] discussed and described, with vast scientific vision, arsenic in groundwater and its poisoning and risk assessment. The author elucidated in details arsenic poisoning through ages, the global scenario of groundwater arsenic catastrophe, spatial mapping, spatial planning, and public participation in confronting this environmental disaster.[4] Environmental concerns and epidemiological and spatial assessment are the areas of scientific endeavor in this paper. Apart from Bangladesh and India,

196   Green Chemistry and Biodiversity: Principles, Techniques, and Correlations

which between them have the largest problem, there have been serious warnings from Argentina, Chile, Taiwan, Vietnam, China, Pakistan, and southwestern parts of the USA. The science and technology pursuits and deep inquiry are thus the need of the hour. This book deals with the methodological issues of spatial, quantitative, and qualitative inquiries of arsenic poisoning, for instance, the use of geographical information systems (GIS) to investigate the distribution of arsenic-laced water in space time.[4]

Chowdhury et al.[5] discussed and described with lucid and cogent insight groundwater arsenic contamination in Bangladesh and West Bengal, India. Today, science and technology need to be re-envisioned and re-envisaged as regards to environmental remediation.[5] Groundwater arsenic poisoning in developing and developed nations around the world is a curse to human mankind. Scientific innovation, deep scientific and technological validation, and the futuristic vision of engineering science will all lead a long and visionary way in the true realization of environmental protection science. Nine districts in West Bengal, India, and 42 districts in Bangladesh have arsenic levels in groundwater above the World Health Organization maximum permissible limit of 50 µg/L.[5] The area and population of the 42 districts in Bangladesh and nine districts in West Bengal are 92,106 km² and 79.9 million and 38,865 km² and 42.7 million, respectively.[5] A concerted effort by engineers, scientists, and the civil society to grapple this disastrous environmental engineering problem is the need of the hour. In Bangladesh, the authors have identified 492 affected villages in 141 police stations of 42 affected districts.[5] The authors have collected 10,991 samples from 42 arsenic-affected districts in Bangladesh for analysis, 58,166 water samples from nine arsenic affected districts in West Bengal, India. Of the water samples analyzed, 59% and 34%, respectively, contained arsenic levels above 50 µg/L. This is a veritable scientific catastrophe.[5] The authors surveyed 27 of the 42 districts in Bangladesh for arsenic patients; they identified patients with arsenic skin lesions in 25 districts. In West Bengal, they identified patients with lesions in seven of the nine districts.[5] After 10 years of study in Bangladesh, according to the authors, it is just the tip of a catastrophic iceberg. Human civilization's immense scientific prowess, girth, and perseverance are at a devastation as arsenic groundwater poisoning destroys the global scientific firmament.[5] To combat the arsenic crisis, the authors described the following steps: (1) In most villages surveyed in Bangladesh and West Bengal, an average of 35% of tube wells contain water that is safe to drink,

Integrated Water Resource Management and Nanotechnology Applications   197

the safe tube wells should be tested for arsenic every 5–6 months.[5] (2) Epidemiological research is immensely needed in the arsenic-affected areas of West Bengal and Bangladesh to characterize and quantify the arsenic-related public health issues and to vastly document the long-term health benefits of the teeming millions in developing countries around the world. (3) In Bangladesh and West Bengal, the surface water resources of sweet water such as rivers, wetlands, and flooded river basins are among the largest in the world. These two vastly delta areas are known as the land of rivers and have approximately 2000 mm annual rainfall. Watershed management and integrated water resource management are the necessities of human scientific innovation today. In this case, the authors of this report deeply suggested these approaches.[5] (4) Today, it is widely accepted that there is no medicine for chronic arsenic toxicity; safe water, nutritious food, and physical exercise are the only prevented measures toward a sound health system. Thus the authors stressed on the immense needs of re-envisioning and revamping of the environmental engineering scenario of arsenic groundwater remediation and environmental remediation. This entire treatise is a veritable eye-opener toward the needs of the scientific community and the civil society in grappling this unending human crisis. This effort and this approach of scientific research and the futuristic vision will surely unfold the hidden intricacies of arsenic crisis in Bangladesh, India, and other developing nations of the world.[5]

Chakraborti et al. (2003)[6] discussed with immense scientific and technological concern arsenic groundwater contamination in the middle Ganga plain , Bihar, India, and the ever-growing future danger. The alarming pandemic of arsenic contamination in groundwater and drinking water in West Bengal, India, and all of Bangladesh has been thought to be limited to the entire Ganges delta (the lower Ganga plain), despite early survey reports of arsenic contamination in groundwater in the Union territory of Chandigarh and its surroundings in the northwestern Upper Ganga plain and recent scientific findings in the Terai area of Nepal.[6] Vast scientific vision, scientific forbearance, and the immense scientific discernment are the necessities of research pursuit and research innovation globally. The authors, in this treatise, deeply discussed the vast geographical domain of arsenic poisoning in South Asia and the immense health issues related to it. Human scientific vision and vast scientific profundity in arsenic groundwater research will veritably unravel the intricacies and scientific barriers in grappling the world's largest environmental crisis.[6]

In this section, the author deeply dealt with arsenic groundwater remediation which stands as a disaster in human scientific progress. Civil society participation and scientific prowess of environmental engineering science are the utmost needs of mankind today. The vast challenges and the vision of chemical process engineering and environmental engineering are deeply threaded in arsenic and heavy metal groundwater remediation research. This vision and the in-depth targets of science are related in this section.

13.11 SIGNIFICANT SCIENTIFIC RESEARCH PURSUIT IN THE FIELD OF NANOTECHNOLOGY AND NANOFILTRATION

Nanofiltration and membrane separation processes are the challenging areas of scientific endeavor in environmental engineering and chemical process engineering in today's modern science. In this section, the author deeply elucidates the scientific need of novel separation processes and nanotechnology applications in environmental engineering, chemical process engineering, and water process engineering. Nanofiltration, ultrafiltration, and microfiltration are the frontier avenues of scientific research in membrane separation processes. The onus and responsibility of global water shortage and the futuristic vision of environmental engineering lie in the hands of scientists, engineers, and humanity. Technology and engineering science of environmental remediation are in the crucial juncture of vision and deep scientific sagacity. In this section, the author deeply comprehends the utmost need of human society—the provision of clean drinking water and the greater emancipation and mitigation of global water shortage. Membrane separation processes can play a remarkable role in mitigating water purification and environmental engineering separation processes.

Cheryan[7] discussed with deep scientific conscience in a remarkable treatise ultrafiltration and microfiltration. Technological vision, engineering profundity, and the needs of humanity are the torchbearers toward a newer scientific regeneration in the field of novel separation processes, chemical process engineering, and environmental engineering science.[7] Ultrafiltration and microfiltration are the major areas of membrane separation processes. The author discussed with cogent insight membrane chemistry, structure and function, membrane properties, performance and engineering models, fouling and cleaning, process design, and the vast area of applications of ultrafiltration and microfiltration.[7] The development of Sourirajan-Loeb synthetic membrane in 1960 provided a valuable and a remarkable tool to the process industries and it faced tremendous scientific barriers in its early days. Technology and engineering science developed immensely fast after those initial experiments.[7] The situation is different today: membranes are exceedingly robust, modules and equipment are better designed and there is a greater scientific understanding and a greater scientific vision in the field of fouling phenomenon.[7] Original equipment manufacturers and the better scientific understanding of process design are the forerunners of membrane separation processes today. The primary and important role of a membrane is to act as a selective barrier. It should permit passage of certain components and retain other components of a mixture of organic and inorganic compounds.[7] Membranes can be classified by (1) nature of the membrane—natural versus synthetic; (2) structure of the membrane—porous versus nonporous; (3) application of the membrane—gaseous phase separations, gas–liquid, liquid–liquid, etc; (4) mechanism of membrane action—adsorptive versus diffusive, ion-exchange, osmotic, and non-selective membranes.[7] Nanofiltration is a relatively new process that uses charged membranes with pores larger that reverse osmosis membranes. When first developed in the 1960s, ultrafiltration and reverse osmosis and later joined by their sister pressure-driven processes, microfiltration, and nanofiltration constituted the first continuous molecular separation processes that do not involve a interphase mass transfer operation.[7] A further definite advantage of membrane technology, as compared to conventional dewatering process, is the absence of a change in phase or state of the solvent.[7] Energy saving aspects should be taken into account when choosing membrane separation processes over other unit operations of chemical engineering. The general methods of manufacture of membranes are dealt with in minute details in this well-researched treatise. The mathematical models in the mass transfer operations in membranes are equally discussed in details and with large scientific conscience.[7] Membrane separation processes are the fountainhead of scientific endeavor in unit operations of chemical engineering. Technological advancements and enriched scientific and engineering achievements in environmental engineering will surely open up newer avenues in membrane science and process engineering design as regards novel separation processes.[7]

The Royal Society and The Royal Academy of Engineering Report (2004)[8] discussed with vast vision and scientific conscience opportunities and uncertainties in nanoscience and nanotechnologies. The authors of this report discussed with immense scientific zeal science and applications of nanoscience and nanotechnologies, nanomanufacturing, and industrial application of nanotechnologies, possible adverse health effects, social and ethical issues, and regulatory issues.[8] Today is the era of multidisciplinary areas of scientific research pursuit. Renewable energy research should be the major pillar of nanotechnology and nano-engineering.[8] The theme of this study was to define: (1) what is meant by nanoscience and nanotechnologies, (2) summarize the current state of scientific discernment in the field of nanotechnologies, (3) identify the specific applications of the new technologies, (4) carry out a forward look to see how the technologies might be used in future, (5) identify the health and safety aspects of nano-science and nanotechnology, and (6) identify and envision areas where regulation are vastly needed.[8] Current and potential areas of nanotechnologies are nanomaterials, metrology, electronics, optoelectronics and information and computer technology, bionanotechnology, and nanomedicine. Green nanotechnology is the other large pillar of research pursuit in engineering today.[8] Science and technology have really advanced very fast over the last few decades and so also nanotechnology and nano-engineering.[8] Regulatory aspects and health aspects of nanotechnology are immensely important as human civilization trudges forward toward a newer era.[8] This entire report targets the scientific vision of today's nanotechnology and nano-engineering and veritably opens up newer dimensions in knowledge and applications of nanoscience to human civilization in decades to come.

Federal Ministry of Education and Research Report, Germany (2016)[9] deeply dealt with action plan for nanotechnology for 2020. This is a high-tech strategy report on the part of German Government. This is also an inter-departmental strategy of the Federal Government.[9] This report deeply discusses digital economy and the society, sustainable economy and energy, innovative working environment, healthy living, intelligent mobility, and civil security as contribution to the future tasks of the high-tech strategy. In the field of nanotechnology applications, creating the framework conditions for sustainable innovations are the other major pillars of this treatise.[9] Nanotechnology is playing an important role in our lives as science and technology surge forward.[9] Applications based on nanotechnology have become more significant economically in the

Integrated Water Resource Management and Nanotechnology Applications   201

past few years and are surpassing vast and versatile scientific boundaries. Nanotechnology-based product innovations play an increasingly important role in many areas of life such as health and nutrition, the workplace, mobility, and energy production. Renewable technology applications and energy sustainability are the important and remarkable areas of scientific endeavor in Federal Republic of Germany.[9] Thus the immediate necessity of this strategic report. Nanotechnology deals with the controlled manufacture and use of nanomaterials and components with functionally relevant structures sizes below 100 nm in at least one dimension.[9] In Germany, a more stringent and concise innovation policy is necessary which addresses both opportunity safeguarding and risk scientific research pursuit. Human scientific progress in this nation today is in the path of newer regeneration and newer scientific ingenuity. This report is a global eye-opener toward a global consensus in the field of nanotechnology applications and nano-engineering emancipation.[9]

Water purification, environmental engineering separation processes, and nanotechnology are integrated together in the future scientific emancipation globally. In this treatise, the author deeply deals with the success of nano-engineering and nanotechnology in diverse applications of science and engineering. Nanomaterials and engineered nanomaterials are changing the pillars of research pursuit in nanotechnology today. Thus the scientific history of human civilization will be more re-organized and re-envisioned if nanomaterials and engineered nanomaterials applications to humanity are more enhanced.

13.12   SIGNIFICANT SCIENTIFIC ADVANCES IN INTEGRATED WATER RESOURCE MANAGEMENT

Scientific advances in integrated water resource management are in the scientific phase of immense challenges and vast vision. Human civilization's immense scientific prowess, girth, and determination will surely lead a visionary way toward a newer scientific revival of environmental engineering and chemical process engineering. In this section, the author deeply pronounces the success of engineering science and technology in nanotechnology applications in environmental engineering and integrated water resource management. Today, human civilization faces the worst environmental engineering disaster that is arsenic groundwater contamination in many developing and developed nations around the world. This area of scientific research disaster needs to be mitigated with war-footing as science and engineering move forward.

Global Water Partnership Report [10] discussed with immense scientific conscience integrated water resources management in basins. Water resource engineering and water resource management are the culmination of vast and versatile scientific research pursuit globally today. Global Water Partnership is an international network whose vast vision is for a water secure world. The GWP mission targets sustainable development and management of water resources at all levels. Human scientific vision and human scientific ingenuity are at its scientific and engineering helm as civilization and science move forward.[10] This report is an eye-opener toward the needs of water resources and environmental engineering toward the greater emancipation of science and technology. A coordinated development and management of water, land, and related resources by maximizing economic and social welfare is the need of integrated water resource management globally.[10] The International Network of Basin Organizations established in 1994 is an international network that highly supports the implementation of integrated water resources management in river and lake basins and lake aquifers.[10] It largely supports of basin management globally along with sustainable development and environmental management.[10] The areas of research pursuit discussed in this report are basin management, political approach and basin management systems, law and policy, water management framework, strategic long term planning, basic action plans, basin information systems and monitoring, and the vast needed awareness of the water resource management needs.[10] The success of science and engineering of water resource management and the related wide world of environmental engineering science are deeply investigated in this report.[10] A new awakening in the field of environmental science has ushered in globally today as integrated water resource management enters a newer scientific age.

Gumbo et al. (2001)[11] discussed with lucid and cogent insight principles of integrated water resources management. The world's freshwater resources are slowly decreasing. Thus, the need of science and engineering forays and unending scientific vision in the field of water resource engineering.[11] Growth in population, vast increase in economic activity, and largely improved standards of living of mankind lead to increased competition and conflicts of water resources. Here comes the importance

Integrated Water Resource Management and Nanotechnology Applications 203

of water resource management.[11] There is a global need of comprehensive water resource management because freshwater resources are limited, limited freshwater resources are limited, there is a greater need of safe water resources, techniques are needed to address the competing needs and the vast needs of human society, and there are needs of an intimate relationship between groundwater and surface water.[11] Here comes the scientific vision and deep ingenuity in integrated water resource management. Integrated water resource management then seeks to manage the water resources in a comprehensive, a viable, and a holistic way. Integrated water resource management thus targets the entire water cycle with all its natural aspects and it also vastly acknowledges the interests of the water users in the different sections of human society.[11] Current thinking on the crucial strategic issues is heavily influenced by the so-called Dublin principle, which was formulated during the International Conference on Water and Environment in Dublin in 1992. These principles on integrated water resource management are elucidated in deep details in this treatise.[11]

Scientific advances in the field of integrated water resource management are challenging and visionary for both developing as well as developed countries around the globe. Water resource engineering and environmental engineering are the veritable needs of humanity today. The scientific impasse of water issues needs to be thus envisioned and re-organized as science moves forward.[12–14]

13.13 SCIENTIFIC RESEARCH PURSUIT IN THE FIELD OF GROUNDWATER REMEDIATION

Groundwater remediation and environmental protection are the utmost needs of the hour. The world of challenges, the scientific barriers, and the scientific intricacies in the field of water resource engineering and water resource management are the torchbearers toward a newer scientific thought and a new scientific vision in science and technology globally. Nanotechnology, nanoscience, and nanoengineering are the fountainheads of scientific research pursuit globally. The veritable importance of nanotechnology applications in environmental engineering needs no explanations. The technology and engineering science of nanotechnology in today's world need to be re-organized as civilization moves forward. Nano-particles and engineered nanoparticles have health effects. Those areas of endeavor need to be re-envisioned and re-envisaged. Arsenic and heavy metal groundwater poisoning of drinking water are the curse and barrier of human civilization and human scientific progress today. Here comes the immense importance of nanotechnology and nano-engineering in environmental engineering and diverse branches of chemical process engineering. The scientific status of chemical process engineering and petroleum engineering stands in the midst of forbearance and vision. Thus, these areas needs to be overhauled as mankind surges forward.[12–14]

Qu et al.[15] deeply discussed with vast vision and ingenuity applications of nanotechnology in water and wastewater treatment. Providing clean and potable water is a vision of humanity today.[15] This treatise covers candidate nanomaterials, properties, and mechanisms that enable the applications, advantages, and limitations as compared to the existing processes, and the immense barriers and research needs for commercialization.[15] The authors deeply discussed current and potential applications for water and wastewater treatment mainly carbon-based nano-adsorbents, metal-based nanoadsorbents, polymeric nanoadsorbents, and nanofiber membranes. Thin film nanocomposite membranes and biologically inspired membranes and photocatalysis are the other pillars of this comprehensive treatise.[15]

Science and engineering are immensely thought-provoking as regards applications of nanomaterials in human society. In this entire chapter, the authors deeply pronounce the necessities and the utmost vision of groundwater remediation, environmental remediation, and nanotechnology.

13.14 FUTURE RESEARCH DIRECTIONS AND FUTURE OF SCIENTIFIC THOUGHTS

Future of human civilization and human scientific endeavor is vastly dismal as well as thought-provoking. In this section, the author comprehends the immense future directions in research in the field of integrated water resource management, water purification, and nanotechnology. Future scientific thoughts should target a better scientific model of integrated water resource management and the success of nanotechnology applications in water purification. Technology and engineering of water pollution control and water resource engineering are highly challenged as science and civilization move forward. Scientific inquiry, deep scientific wisdom, and scientific discernment of interdisciplinary science will surely lead a long and effective way in unlocking the truths and intricacies of water resource engineering and water resource management. A vast

Integrated Water Resource Management and Nanotechnology Applications   205

interdisciplinary approach is necessary in the proliferation of environmental engineering science and nanotechnology. This chapter unfurls the needs of human society such as sustainability and renewable energy. The challenge and the vision are brought forward by the scientists and the civil society. Future research directions and future of scientific thoughts should be directed toward more scientific emancipation in the field of nanotechnology applications in diverse areas of science and engineering which includes environmental protection. The scientific status of environment is immensely dismal in modern day human civilization. In this treatise, the author deeply comprehends with vast scientific insight the needs of humanity and the success of sustainability applications in human scientific progress. Today is the world of interdisciplinary science and engineering. Integrated water resource management encompasses diverse areas of scientific endeavor such as environmental engineering and water resource engineering. The author deeply elucidates with lucidity the needs of scientific inquiry in sustainability application in human society. The challenge, the vision, and the targets need to envisioned by the scientists and the civil society as science moves forward.

13.15 CONCLUSION, SUMMARY, AND FUTURE ENVIRONMENTAL PERSPECTIVES

Environmental perspectives need to be reorganized and revamped in today's scientific world. Environmental engineering and chemical process engineering are the needs of humanity and the marvels of science and technology today. This treatise definitely opens up newer visionary areas and newer thoughts in the field of water resource engineering and water resource management. Technology and engineering science stand retrogressive as environmental concerns and fossil fuel concerns destroy the vast scientific fabric. Science has progressed a lot today yet in some instances a major failure. Here comes the need of newer visionary areas such as renewable energy, water pollution control, water resource engineering, and integrated water resource management. The contributions of human factor engineering and systems engineering to human scientific progress are immense and ground-breaking. Today, human factor engineering needs to be linked with integrated water resource management and environmental engineering. This will be the future vision of science and technology. Water purification and the status of environment is immensely

206 Green Chemistry and Biodiversity: Principles, Techniques, and Correlations

dismal. This chapter veritably opens up new visionary thoughts in the field of also nanotechnology approaches in environmental engineering. Technology has progressed at a rapid pace in today's world and energy sustainability and energy security are the utmost needs of the hour. This treatise will surely be an eye-opener toward a new era of integrated water resource management and the world of nanotechnology. The question of environmental sustainability is highly daunting and far-reaching in today's human civilization. The authors deeply stressed the need of sustainable development in the progress of human scientific endeavor. Technology and engineering science of renewable energy will then surely open up new vistas in the field of energy sustainability. Human scientific regeneration in sustainable development will then usher in a new era of scientific hope and scientific determination.

KEYWORDS

- **nanotechnology**
- **purification**
- **sustainability**
- **vision**
- **water**
- **water resource**

REFERENCES

1. Kanchi. S. Nanotechnology for Water Treatment. *J. Environ. Anal. Chem.* **2014,** *1* (2), 1–3.
2. Sengupta, B.; Chatterjee, S.; Rott, U.; Kauffman, H.; Bandopadhyay, A.; DeGroot, W.; Nag, N. K.; Carbonell-Barrachina, A. A.; Mukhopadhyay, S. A simple Chemical Free Arsenic Removal Method for Community Water Supply—A Case Study for West Bengal, India. *Environ. Poll.* **2009,** *157* (12), 3351–3353.
3. Hashim, M. A.; Mukhopadhyay, S.; Sahu, J. N.; Sengupta, B. Remediation Technologies for Heavy Metal Contaminated Groundwater. *J. Environ. Manag.* **2011,** *92*, 2355–2388.
4. Hassan, M. M. *Arsenic in Groundwater: Poisoning and Risk Aassessment;* CRC Press: Boca Raton, USA, 2018.

Integrated Water Resource Management and Nanotechnology Applications   207

5. Choudhury, U. K.; Biswas, B. K.; Choudhury, T. R.; Samanta, G.; Mandal, B. K.; Basu., G. K.; Chanda, C. R.; Lodh, D.; Saha, K. C.; Mukherjee, S. K.; Roy, S.; Kabir, S.; Quamruzzaman, Q.; Chakraborti, D. Groundwater Arsenic Contamination in Bangladesh and West Bengal, India. *Environ. Health Persp.* **2000,** *108* (5), 393–397.

6. Chakraborti, D.; Mukherjee, S. K.; Pati, S.; Sengupta, M. K.; Rahman, M. M.; Chowdhury, U. K.; Lodh, D.; Chanda, C. R.; Chakraborti, A. K.; Basu, G. K. Arsenic Groundwater Contamination in Middle Ganga Plain, Bihar, India: A Future Danger? *Environ. Health Pers.* **2003,** *111* (9), 1194–1201.

7. Cheryan, M. *Ultrafiltration and Microfiltration Handbook*; Technomic Publishing Company Inc.: Lancaster, Pennsylvania, USA, 1998.

8. The Royal Society Report, United Kingdon. *Nanoscience and Nanotechnologies: Opportunities and Uncertainties*; Royal Society & The Royal Academy of Engineering: London, United Kingdom, 2004.

9. Federal Ministry of Education and Research Report, Federal Republic of Germany. *Action Plan Nanotechnology (2020), An Inter-departmental Strategy of the Federal Government*; German Government: Germany, 2016.

10. Global Water Partnership Report. *A Handbook for Integrated Water Resources Management in Basins*; International Water Association: London, United Kingdom, 2009.

11. Gumbo, B.; van der Zaag, P. Principles of Water Resources Management, Global Water Partnership Southern Africa, Southern Africa Youth Forum, 24–25 September, 2001, Harare, Zimbabwe.

12. Barrow, C. J. *Environmental Management and Development*; Routledge, Taylor and Francis Group: USA, 2005.

13. Barrow, C. J. *Environmental Management for Sustainable Development*; Routledge, Taylor and Francis Group: USA, 2006.

14. Shannon, M. A.; Bohn, P. W.; Elimelech, M.; Georgiadis, J. G.; Marinas, B. J.; Mayes, A. M. *Science and Technology for Water Purification in the Coming Decades*; Nature Publishing Group: USA, 2008; pp 301–310.

15. Qu, X.; Alvarez, P. J. J.; Li, Q. Applications of Nanotechnology in Water and Wastewater Treatment. *Water Res.* **2013,** *47* (12), 3931–3946.

CHAPTER 14

Precision Personalized Medicine from Theory to Practice: Cancer

FRANCISCO TORRENS[1*] and GLORIA CASTELLANO[2]

[1]*Institut Universitari de Ciència Molecular, Universitat de València, Edifici d'Instituts de Paterna, P. O. Box 22085, E46071 València, Spain*

[2]*Departamento de Ciencias Experimentales y Matemáticas, Facultad de Veterinaria y Ciencias Experimentales, Universidad Católica de Valencia San Vicente Mártir, Guillem de Castro-94, E46001 València, Spain*

Corresponding author. E-mail: torrens@uv.es

ABSTRACT

Precision personalized medicine means a change of paradigm. Intrinsically, every tumor–host complex is a rare disease. Breast cancer is not a unique disease but a group of tumor subtypes with different natural history. The traditional model is exhausted and does not serve people. Why is there no cure yet against cancer? Cancer is not a unique disease. Cancer cannot be treated as an infectious disease. To design therapies that destroy viruses or bacteria without affecting people's own cells results simpler. The problem is that tumor cells are cells of people's own body. Precision oncology increases the rates of survival. Precision oncology allows safer treatments directed vs. targets (decreases AEs). Precision oncology removes the use of unnecessary or ineffective treatments. Precision oncology improves system efficiency. A critical moment in precision medicine: Information is more important than drugs. Patients and healthcares search for results. Academics must lead to change. Advantage: They dispose of clinical and scientific knowledge.

14.1 INTRODUCTION

Setting the scene: Precision personalized medicine (PPM) from theory to practice. What can people learn? The PPM means a change of paradigm. Intrinsically, every tumor–host complex is a rare disease. Breast cancer (BC) is not a unique disease but a group of tumor subtypes with different natural history. The traditional model is exhausted and does not serve people.

Why is there no cure yet against cancer? Cancer is not a unique disease. Cancer cannot be treated as an infectious disease. To design therapies that destroy viruses or bacteria without affecting people's own cells results simpler. The problem is that tumor cells are cells of people's own body. The PPM oncology (PO) increases the rates of survival. The PO allows safer treatments directed *vs.* targets (decreases AEs). The PO removes the use of unnecessary or ineffective treatments. The PO improves system efficiency. A critical moment in PPM: Information is more important than drugs. Patients and healthcares search for results. Academics must lead to change. Advantage: They dispose of clinical and scientific knowledge.

In earlier publications it was informed the modeling of complex multi-cellular systems and tumor–immune cells competition,[1] information theoretic entropy for molecular classification of oxadiazolamines as potential therapeutic agents,[2] molecular classification of 5-amino-2-aroylquinolines and 4-aroyl-6,7,8-trimethoxyquinolines as highly potent tubulin polymerization inhibitors,[3] polyphenolic phytochemicals in cancer prevention, therapy, bioavailability versus bioefficacy,[4] molecular classification of antitubulin agents with indole ring binding at colchicine-binding site,[5] molecular classification of 2-phenylindole-3-carbaldehydes as potential antimitotic agents in human breast cancer cells,[6] cancer and its hypotheses.[7] It was reported how human immunodeficiency virus/acquired immunodeficiency syndrome destroy immune defenses, hypothesis,[8] 2014 emergence, spread, uncontrolled Ebola outbreak,[9,10] and Ebola virus disease, questions, ideas, hypotheses and models.[11] The present report reviews PPM from theory to practice, from molecular biology to medicine, present, future, clinical research, research in a comic style and cancer (especially, BC). The aim of this work is to initiate a debate by suggesting a number of questions (Q), which can arise when addressing subjects of PPM and hypotheses on cancer, in different fields, and providing, when possible, answers (A) and hypotheses (H).

Personalized Medicine for Cancer

14.2 FROM MOLECULAR BIOLOGY TO MEDICINE: FOUNDATIONS OF PPM

Lluch organized course PPM.[12] A hypothesis was proposed.
H1. (Martín de Dios). PPM means a change of paradigm.
Pérez-Alonso proposed Q/A/H on genetics foundations of PPM and hereditary cancer.[13]

Q1. (Tolosa, 2018). How Many Genes has Human Genome?[14]

A1. *Ca.* 21 000 genes.

H2. Expanded central dogma (CD) (ECD).[15]

H3. CD: DNA–Transcription→Ribonucleic acid (RNA)–Translation→Protein.

H4. ECD: Transcriptional regulation→RNA processing/translational regulation→Posttranslational modification.

H5. Heredity types: simple Mendelian (monogenic diseases); complex (multifactorial diseases).

H6. (Hamburg and Collins, 2010). The Path to PPM.[16]

Q2. (Hamburg and Collins, 2010). Which genetic markers have the most clinical significance?

Q3. (FDA). The revised prescription drug labeling, why were the formal changes necessary?

H7. (Wang, McLeod, and Weinshilboum, 2011). Genomics and Drug Response.[17]

H8. (Goldberger and Buxton, 2013). PPM versus Guideline-based Medicine.[18]

Q4. (Goldberger and Buxton, 2013). Where were patients recruited?

Q5. (President's Council of Advisors on Science and Technology). Who will benefit?

Q6. (President's Council of Advisors on Science and Technology). Who will not benefit?

H9. (Goldberger, 2013). Subpopulations in a trial cohort are unlikely to benefit from intervention.

H10. (McPherson, 2014). A Defining Decade in DNA Sequencing.[19]

212 Green Chemistry and Biodiversity: Principles, Techniques, and Correlations

H11. (Biesecker and Green, 2014). Diagnostic Clinical Genome and Exome Sequencing.[20]

H12. (Biesecker, 2014). Clinicians understand clinical-genome/exome-sequencing (CGES) diagnoses.

Q7. (Biesecker and Green, 2014). How to select the patients most likely to benefit from CGES?

Q8. (Biesecker and Green, 2014). How clinicians should order such testing?

Q9. (Biesecker and Green, 2014). How clinicians should interpret the results?

Q10. (Biesecker and Green, 2014). How clinicians should communicate the results to their patients?

H13. (Biesecker, 2014). Exome-sequencing (ES) advantages: Higher exons coverage; costs less; offered by more laboratories.

H14. (Biesecker, 2014). Genome-sequencing (GS) advantages: More sensitive/accurate detecting structural variation; includes nonexonic regulatory regions; identifies intronic/intergenic-regions common variants.

H15. (Biesecker, 2014). CGES is indicated for detecting rare variants in patients with phenotype suspected due to Mendelian disorder (MD) after known single-gene candidates were eliminated from consideration/with expensive multigene testing.

H16. (Biesecker, 2014). Pretest counseling is important for finding a causative variant/alert patient.

Q11. (Biesecker and Green, 2014). Why can ES be more efficient in a number of clinical scenarios?

H17. (Xue, 2014). MDs molecular diagnostic testing in next-generation sequencing (NGS).[21]

H18. (Xue, Ankala, Wilcox and Hegde, 2014). NGS is changing the paradigm of clinical genetic testing

Q12. (Xue, Ankala, Wilcox and Hegde, 2014). What set of genes is analyzed?

H19. (Xue, 2014). NGS adapted to clinical testing and is changing the paradigm of clinical diagnostics.

Q13. (Xue, 2014). Is NGS the best diagnostic tool in all clinical scenarios?

Q14. (Xue, Ankala, Wilcox and Hegde, 2014). Does targeted single-gene analysis still have a place?

Q15. (Xue, Ankala, Wilcox and Hegde, 2014). Is including more genes always better?

Q16. (Xue, 2014). Why choose gene panel testing when ES gives information on all coding regions?

Q17. (Xue, 2014). What role does clinician play in patient clinical assessment to choose diagnostic?

Q18. (Xue, 2014). What genetic test is that indicated in people clinical case?

Q19. (Xue, Ankala, Wilcox and Hegde, 2014). What genetic test is that indicated?

H20. (Eng, 2013). Clinical Whole-ES for the Diagnosis of MDs.[22]

H21. (Eng, 2014). Molecular Findings Among Patients Referred for Clinical Whole-ES.[23]

H22. (Topol, 2014). PPM from Prewomb to Tomb.[24]

H23. (Kalokairinou, Howard and Borry, 2014). Changes for Consumer Genomics in EU.[25]

Q20. (Borry, 2014). How May EU in Vitro Diagnostic Medical Device Directive Affect Genetic Tests?

H24. (Rahman, 2014). Realizing the Promise of Cancer Predisposition Genes.[26]

H25. Besides diagnosis, patients need medicines: Translational research.

H26. (Rupaimoole and Slack, 2017). Micro (mi)RNA Therapeutics: To a New Era for Cancer .[27]

H27. (Artero, 2018). miR-23b/miR-218 Silencing Increases Muscleblind-like Expression.[28]

Q21. Where to continue to be informed?

A21. Our journal *Genética Médica*.

H28. (Tolosa, 2018). Gene Breast Cancer (BRCA)1 Dissection to Improve Genetic Diagnosis.[29]

H29. (Gallardo, 2018). Molecular Method Developed for Detecting Minimu Tumour-Cells Levels.[30]

Ribas proposed hypotheses, Qs and As on Applications of Genetics in Clinical Practice.[31]

H30. Metastasis is what kills persons.

Q22. Has cancer always been so frequent?

Q23. Is it the same having predisposition to suffer from cancer as having sporadic cancer?

Q24. Are there genetic factors that predispose cancer?

Q25. All or in some?

Q26. Do external risk factors exist that predispose cancer (or some type)?

Q27. Is there the same incidence of cancer in all the world?

Q28. Do men and women suffer from the same types of cancer?

Q29. Does cancer appear at a certain age?

Q30. Has one the same probability of dying/living of cancer according to the type or where he lives?

Q31. What technologies are used for diagnosis?

Q32. How are treatments decided?

Q33. Are all equal?

Q34. What is PPM?

Q35. What are directed drugs?

Q36. Do people find cancer in antiquity?

A36. A case of osteosarcoma was identified in hominid remains of 1.8 million years ago.

H31. Sporadic cancer risk (CR) factors. Environment: Diet; tobacco; infections; obesity; alcohol; genetics.

Personalized Medicine for Cancer

H32. ERFs: processed food; tobacco; endocrine factors; alcohol; solar irradiance; pollution.

Q37. Has cancer always been so frequent?

A37. No.

Q38. Do ERFs exist that predispose cancer (or some type)?

A38. Yes, they are the base of sporadic cancer.

Q39. Is there the same incidence of cancer in all the world?

A39. No.

Q40. Do men and women suffer from the same types of cancer?

A40. No, though CR is greater for women with *BRCA1/2* gene mutations both face cancer-types elevated lifetime chances.

Q41. Does cancer appear at a determined age?

A41. Yes, according to the type of cancer.

Q42. Has one the same probability of dying/living of cancer according to the type or where he lives?

A42. No.

Q43. Hereditary cancer: Do all bearers of a mutation in a causal gene suffer from the disease?

A43. No, a different penetrance exists; once the disease shows, it can have different expressivity.

H33. BC ERFs: fast food; alcohol; hormone replacement therapy; ovarian cancer (OC).

H34. BC internal risk factors: genetics, age, menarche < 12 years and menopause > 55 years.

H35. (Jorde, Carey and Bamshad, 2015). *Medical Genetics.*[32]

Q44. Are all going to develop hereditary BC and OC?

A44. It will depend.

Q45. Example: What is a pedigree of family (*cf.* Fig. 14.1)?

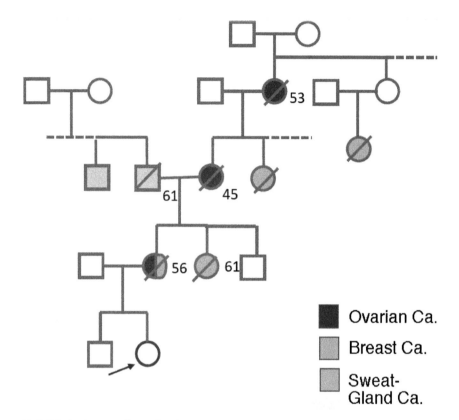

FIGURE 14.1 Angelina Jolie Pitt's pedigree of family. Ca., cancer. *Source:* Ref. [31].

H36. (Kluger and Park, 2013) The Angelina effect (*cf.* Fig. 14.2).[33]
Q46. What will happen to me?
Q47. Will it come back to me?
A47. There is an emotional component in BC.
Q48. (Martín de Dios). What percentage of women are diagnosed not *BRCA1*, not *BRCA2*?
A48. 50% not *BRCA1*, not *BRCA2*; 25% *BRCA1*; 25% *BRCA2*.

Personalized Medicine for Cancer

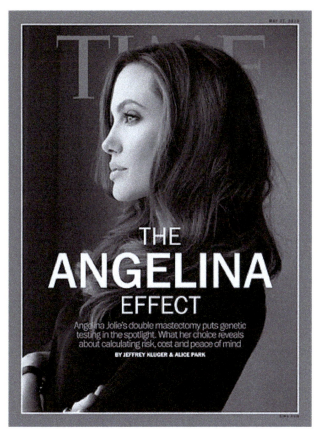

FIGURE 14.2 The Angelina Effect.

Jantús–Lewintre proposed questions, answers and Hs on PPM in clinical practice.[34]

Q49. PPM: Why…?
Q50. What…?
Q51. Where…?
Q52. How…?
Q53. Who…?
Q54. PPM: Why?

H37. Oncology. Way to PPM: One-size-fit-all Medicine –From→ Stratified Medicine –To→ PPM.

H38. PPM ensures delivery of the right intervention to the right patient at the right time.

H39. PO: Using molecular information about a patient's cancer to inform treatment.

Q55. PO: What?

A55. Biomarker (BM): Characteristic measurable as an indicator of normal/pathological bioprocess/pharmacological response to therapy.

Q56. [BM definitions working group (BDWG), 2001]. Am I at increased risk for cancer?[35]

Q57. (BDWG, 2001). Do I have cancer?

Q58. (BDWG, 2001). What type of cancer do I have?

Q59. (BDWG, 2001). What is the expected course of my cancer?

Q60. (BDWG, 2001). Will my cancer respond to this drug?

Q61. (BDWG, 2001). Should I receive a normal or lower dose or no dose?

Q62. (BDWG, 2001). How is my cancer responding to this treatment?

Q63. (BDWG, 2001). Will my cancer come back?

Q64. (BDWG, 2001). Who are likely to benefit?

Q65. (BDWG, 2001). Who are not likely to benefit?

Q66. BMs: What alterations do people search for?

H40. (Hudson, 2013). Tumor characteristics: DNA mutation; copy-number variation; gene expression; DNA methylation; miRNA activity; cellular protein activity.[36]

H41. (Chin and Gray, 2008). Translating Insights from the Cancer Genome into Clinical Practice.[37]

Q67. BMs: What alterations and how many do people search for?

A67. Small number of mountains: genes altered in a high percentage of tumors; a larger number of hills: genes altered infrequently.

Q68. BMs: What alterations do people search for?

Personalized Medicine for Cancer

A68. Single BMs versus signatures.

Q69. PO: Where?

Q70. BMs in cancer: Where to search for?

A70. Sample of tumor tissue.

H42. The paradigm: The tissue is the issue.

H43. Formalin-fixed paraffin-embedded biopsies are standard for molecular tests in PO.

H44. However, in practice, they have some limitations.

H45. Traditional procedure: Obtain tumor tissue block→Manual microdissection→DNA isolation/purification→DNA quantity assessment→Gene-status assessment®Treatment decision.

H46. Necessity of new strategies → liquid biopsy (LB).

H47. (Camps, 2016). Circulating tumor cells vs. (ct)DNA in lung cancer (LC)?[38]

Q71. (Camps, 2016). Circulating tumor cells versus (ct)DNA in LC: Which one will win?

Q72. LB: Why?

A72. It allows in solving the problems of heterogeneity and plasticity (evolutionary changes).

H48. Techniques of great sensitivity are required.

Q73. PO: How do people do the test?

A73. In tissue and LB.

Q74. What type of determination am I going to implement?

A74. In tumor tissue, all DNA is tumoral; in blood, not much DNA is tumoral (ctDNA).

Q75. Which technique?

Q76. Which genes must I analyze?

Q77. How many genes must I analyze?

Q78. Sensitivity?

Q79. NGS: Which type of information does it generate?

220   Green Chemistry and Biodiversity: Principles, Techniques, and Correlations

A79. Targeted sequencing. Predominant applications: point mutations and deletions.

Q80. ctDNA LB: From YES/NO to how much?

Q81. Negative sensitivity?

Q82. Positive sensitivity?

A82. Droplet digital polymerase chain reaction (ddPCR) > quantitative (q)PCR > NGS.

Q83. Model–view–controller, valid?

Q84. PO: Who must carry out these molecular determinations?

Q85. PO?

She provided the following conclusions (Cs) on PO.[39]

C1. PO increases the rates of survival.

C2. PO allows safer treatments directed *vs.* targets (decreases AEs).

C3. PO removes the use of unnecessary or ineffective treatments.

C4. PO improves system efficiency (not to use resources in those that are not going to benefit).

14.3   PRESENT AND FUTURE OF PRECISION PERSONALIZED MEDICINE

Llombart-Cussac proposed Qs, As, and Hs on molecular targets and development of new drugs.[40]

Q1. What is cancer?

A1. Process generated by accumulation of mutations in genetic code in a concrete cell (genomics).

H1. Reductionism in cancer: Way of classifying complex systems.

H2. Reductionism in cancer: Impossible to integrate all mechanisms in a congruent way.

H3. Intrinsically, every tumor–host complex is a rare disease.

H4. Reductionism in cancer: It is impossible to integrate all cases in a unique model.

Personalized Medicine for Cancer 221

H5. Present classification of BC: anatomical and molecular.

H6. Present anatomical classification of BC: Tumor size–lymph nodes–metastasis.

H7. Present molecular classification of BC: phenotype.

Q2. Are these two patients equal?

Q3. Is this PPM or do people go deeply into reductionist medicine?[41]

H8. Synthetic lethality: Tumor alters by BRCA loss but subsists thanks to poly(adenosine diphosphate-ribose polymerase (PARP); if people block the latter, cells die.

Q4. How can people improve predictive power in sensitivity to PARP inhibitors?

Q5. What specific weight has every one of the involved genes?

Q6. Future?

A6. Individualized algorithms and PPM.

H9. (P. Picasso). Computers are useless. They only know to give answers.

H10. (J. Wagensberg). To change the answer is evolution. To change the question is revolution.

H11. Triple-negative BC (TNBC) is similar to microcytic LC.

He provided the following conclusions.

C1. Critical moment in PPM: Information (determinations) is more important than drugs.

C2. Patients and healthcares search for results.

C3. Academics must lead change. (1) Advantage: they dispose clinical/scientific knowledge. (2) Challenges: adequate structures (university/hospitals/spin-off sources/pair collaborations]; cooperative works–transformation; alternatives: research models integrating best of public/private systems [MedSIR]).

H12. (G. Marañón). There are no diseases but patients.

Cervantes proposed Hs and Qs on approach from PPM in an oncology department.[42]

H13. (Dienstmann and Tabernero, 2017). Cancer: A PPM Approach to tumor treatment.[43]

H13. (Picco and Garnett, 2017). A Roadmap for Cancer PPM via PPM Models.[44]

H15. (Low, 2015). An Oncogenic NTRK Fusion in a Patient with Soft-tissue Sarcoma.[45]

Q7. (Low, 2015). *CDKN2A/B* deletion?

H16. (Hyman, 2018). Efficacy of Larotrectinib in TRK Fusion-positive Cancers.[46]

H17. Cancer diagnosis evolution: Clinical diagnosis®Pathology diagnosis→Molecular diagnosis→PO.

H18. Clinical approach→Pathology approach→Molecular approach.

H19. (Garraway, Verwey and Ballman, 2013). Genomics-driven cancer medicine.

H20. (André, 2018). Developing Anticancer Drugs in Orphan Molecular Entities (OMEs).[47]

H21. (André, 2018). OMEs drug development: Rare genomic altera-tions with unmet medical need→Single-group practice-changing trial→After regulatory approval.

Q8. (André, 2018). Pending issues: How to define an OME in oncology?

Q9. (André, 2018). Incidence: Orphan Drug Act?

H22. (Soria, 2017). High-throughput Genomics and Clinical Outcome in Advanced Cancers.[48]

H23. (Bose, 2015). HER2 Mutations are Targets for Colorectal Cancer (CRC) Treatment.[149]

H24. (Siena, 2016). Trastuzumab/Lapatinib in HERACLES Treatment and Assessments.[50]

H25. (Siena, 2015). Representative Contrast-enhanced Computed Tomography Scans of 2 Responders.

H26. (Rubin, 2017). Personalized In Vitro and In Vivo Cancer Models to Guide PPM.[51]

H27. (André, 2018). A framework to rank genomic alterations as targets for cancer PPM.

Personalized Medicine for Cancer

H28. (Valeri, 2018). Patient-derived Organoids Model Treatment Response.[52]

Lluch proposed hypotheses, questions and As on approach from PPM in breast cancer.[53]

H29. Problem of public health (PH).

H30. Therapy is far more individualized based on disease characteristics/ stage/patient preference.

H31. (Rodriguez, Walters and Burke, 2013). 60 Years of Survival Outcomes of Cancer.[54]

H32. (Walgren, Meucci, and McLeod, 2005). Notpersonalized treatment.[55]

Q10. (Walgren, Meucci, and McLeod, 2005). Will the real genes please stand up?

H33. (Hanahan and Weinberg, 2000). *The Hallmarks of Cancer*.[56]

H34. Most malign tumors require dysregulation of one/some processes to grow/invade/metastatize.

H35. BC is not a unique disease but a group of tumor subtypes with different natural history.

Q11. What is PPM?

A11. PPM is a model that pretends to fit every treatment via molecular or genomic tools.

H36. (Weigelt and Reis-Filho, 2009). Histological and Molecular Types of BC.[57]

Q12. (Weigelt and Reis-Filho, 2009). Is there a unifying taxonomy?

H37. Three biological subtypes of BC: ER^+ (65–75%), $HER2^+$ (15–20%) and TNBC (15%).

H38. (EBCTCG, 2011). BC Hormone Receptors Relevance to Adjuvant Tamoxifen Efficacy.[58]

She provided the following conclusions on hormone therapy (HT).

C4. Luminal disease ER^+ presents a clinical benefit to therapies directed vs. proper receptor.

C5. In initial stages results improvement comes from young women– HTs duration in menopausal women combination.

Q13. Intermediate risk: either HT/benefit or benefit of chemotherapy (CT)?

She provided conclusions on genetic platforms (GPs) utility in treatment selection.

C6. GPs have unequivocal prognostic value in patients with luminal BC.

C7. GPs are tool useful/necessary in making decisions on adjuvant therapy (AT) in negative ganglia.

C8. Controversy exists on GPs utility in making decisions on AT in *positive ganglia*/other patients.

C9. Prospective clinical studies exist that will provide information on points with controversy.

She provided the following additional hypotheses, questions, and answer.

H39. (Tabernero, 2013). Development of phosphoinositide-3-kinase (PI3K) inhibitors.[59]

H40. (Fruman and Rommel, 2014). PI3K and Cancer: Lessons, Challenges, and Opportunities.[60]

H41. (Foulkes, Smith and Reis-Filho, 2010). Current Concepts: TNBC.[61]

H42. (Aarts, Linardopoulos and Turner, 2013). Cell-cycle-kinases Tumor Selective Targeting .[62]

H43. (Clemons, 2009). Survival outcomes for patients with metastatic TNBC.[63]

H44. (Narod, 2007). TNBC: Clinical Features and Patterns of Recurrence.[64]

H45. TNBC paradox: Chemosensitive… but relapse aggressive with worse overall survival (OS).

H46. (Yarden and Sliwkowski, 2001). Untangling the ErbB signalling network.[65]

Q14. How does the epidermal growth factor receptor (EGFR) act in a pathological way?

H47. (DeVita, Jr., Lawrence and Rosenberg, 2014). Cancer: Principles & Practice of Oncology.[66]

Personalized Medicine for Cancer 225

Q15. Immunotoxin?

Q16. How to improve Trastuzumab results in adjuvancy?

A16. Pertuzumab.

H48. (Baselga, 2017), Adjuvant Pertuzumab and Trastuzumab in early HER2⁺ BC.[67]

She provided the following conclusions on AT.

C10. Anti*HER2* treatment changed the natural history of BC.

C11. Anti*HER2* AT (in conjunction with CT) reduces 50% relapses and deaths.

C12. The optimal duration of therapy with Trastuzumab is one year.

C13. Precocious tumors cannot be treated with not much toxic guides.

C14. Pertuzumab adds benefit in tumors with axillary affectation and hormone receptor (HR)⁻.

H49. (Valagussa, 2012). Neoadjuvant Pertuzumab/Trastuzumab Efficacy/Safety in HER2⁺ BC.[68]

She provided the following conclusions on PO.

C15. The era of treating cancer in an equal manner for all has finished.

C16. Understanding cancer molecular bases allows: targets identification; therapeutic focuses.

C17. BMs identification made possible patients selection for different treatments.

C18. The present challenge in oncology is to individualize treatment for every patient and tumor.

14.4 CLINICAL RESEARCH IN PRECISION PERSONALIZED MEDICINE

Berrocal proposed Hs and Qs on bases of immunotherapy (IT) and their application in PPM [69].

H1. (Dranoff, 2004). Cytokines in cancer pathogenesis and cancer therapy.[70]

H2. (Male, Brostoff, Roth and Roitt, 2012). Immunology [71].

H3. (Kahan, 2003). Individuality: The Barrier to Optimal Immunosuppression.[72]

H4. Theory of the three signals.

H5. (Chen and Flies, 2013). Molecular Mechanisms of T-cell Costimulation and Coinhibition.[73]

H6. (Coomes, Pelly and Wilson, 2013). Plasticity Within the $\alpha\beta^+$CD4$^+$ T-cell Lineage.[74]

Q1. (Coomes, Pelly and Wilson, 2013). Plasticity in $\alpha\beta^+$CD4$^+$ T-cell: When, how and what for?

H7. (Dunn, Old, and Schreiber, 2004). Cancer Immunosurveillance/ Editing Immunology.[75]

H8. (Vesely, Kershaw, Schreiber and Smyth, 2011). Cancer Natural Innate/Adaptive Immunity.[76]

H9. (Hanahan and Weinberg, 2011). Hallmarks of cancer: The next generation.[77]

H10. (Hodi, 2009). Guidelines for the Evaluation of Immune Therapy Activity in Solid Tumors.[78]

Q2. Can patients be selected in IT?

H11. (GEM, 2017). Predictive factors of response to IT.[79]

H12. (De Mello, 2017). Programmed Death (PD)-Ligand (L)1 in Nonsmall-cell Lung Cancer (NSCLC).[80]

H13. (Chan, 2014). Response to Cytotoxic T-lymphocyte Antigen (CTLA)-4 Blockade.[81]

H14. (Hellmann, 2018). Genetic features of response to combination IT in NSCLC.[82]

H15. (Rosemberg, 2008). Adoptive Cell Transfer: A Clinical Path to Effective Cancer IT.[83]

H16. (Rosemberg and Restifo, 2015). Adoptive cell transfer as PPN IT for human cancer.[84]

H17. (Ping, Liu, and Zhang, 2017). T-cell Receptor-Engineered T-cells for Cancer Treatment.[85]

H18. (Overwijk, 2013). Mining the Mutanome: Developing Highly PPM ITs.[86]

Personalized Medicine for Cancer 227

Pérez-Fidalgo proposed Qs, A and Hs on PPM in approaching gynecologic cancer.[87]

Q3.   What is a BM?

A3.   (NIH BMs Definitions Working Group, 1998). Characteristic measured as biological/pathogenical/pharmacological-process indicator in reply to therapy.

Q4.   Who will respond to treatment?

Q5.   Who will not respond to treatment?

H19.   (Cancer Genome Atlas Research Net, CGARN, 2011). Integrated OC genomic analyses.[88]

H20.   (Morgan, 1916). *A Critique of the Theory of Evolution.*[89]

H21.   (Bridges, 1922). SL.

Q6.   (Langreth, 2008). *BRCA* in OC. *BRCA* not mutated: Not to treat?

H22.   (Kohn, 2018). Cell cycle checkpoint kinase-1/2 inhibitor in BRCA wild-type (WT) OC [90].

H23.   (Lheureux and Oza, 2016). Endometrial Cancer.[91]

Q7.   (Lheureux and Oza, 2016). Endometrial Cancer: Targeted Therapies Myth or Reality?

Q8.   Prognostic–predictive subgroup?

Roda Pérez proposed H/Q on PPM in approaching gastric (GC)/ hepatobiliary cancer.[92]

H24.   (CGARN, 2014). Comprehensive molecular characterization of gastric adenocarcinoma.[93]

H25.   (Torre, Bray, Siegel, Ferlay, Lortet-Tieulent, and Jemal, 2015). Global Cancer Statistics.[94]

H26.   (De Angelis, 2014). Cancer Survival in Europe by Country/Age: EUROCARE-5 Results.[95]

Q9.   (ESMO, 2016). OS advanced GC disease: IT?

H27.   (Fan and Chu, 2007). Mismatch Repair (MMR) Repairs Mismatches/Insertions/Deletions.[96]

H28.   (Lordick, 2013). Capecitabine/cisPt with/without Cetuximab for patients with GC.[97]

H29.   (Metzger-Filho, Winer and Krop, 2013). Pertuzumab: Optimizing HER2 Blockade.[98]

228   Green Chemistry and Biodiversity: Principles, Techniques, and Correlations

H30.   (Bang, 2010). Trastuzumab+CT vs. CT for HER2[+] GC/Gastro-oesophageal (GOC) Junction Cancer (GOJC).[99]

H31.   (Lordick and Janjigian, 2016). Clinical Impact of Tumor Biology in GOC Management.[100]

H32.   (Tabernero, 2017). Pertuzumab+Trastuzumab+CT for HER2[+] Metastatic GC/GOJC.[101]

H33.   (Kang, 2017). Trastuzumab Emtansine vs. Taxane for HER2[+] Metastatic GC/GOJC.[102]

H34.   (Ellis, 2002). Antivascular Endothelial Growth Factor (VEGF) receptor (VEGFR)–antiEGFR GC Combination Therapies.[103]

H35.   (Fuchs, 2014). Ramucirumab monotherapy for previously treated advanced GC/GOJC.[104]

H36.   (Wilke, 2014). Ramucirumab + Paclitaxel vs. Placebo + Paclitaxel in GC/GOJC.[105]

H37.   (Muro, 2016). Pembrolizumab for Patients with PD-L1-Positive Advanced GC.[106]

H38.   (Kang, 2018). Molecular Characterization of Clinical Responses to PD-1 Inhibition in GC.[107]

H39.   (CGARN, 2014). Comprehensive Molecular Characterization of GC.[108]

H40.   GC types: Epstein–Barr Virus; microsatellite instability (MSI); genomically stable; chromosomal instability.

She provided the following conclusions.

C1.   GC is a heterogeneous disease.

C2.   New PPM therapies should be based on the molecular classification of every tumor.

C3.   IT could be effective in MSI and EBV GCs.

Camps proposed hypotheses, Qs and As on PPM as a new paradigm in healthcare.[109]

H41.   Two strategic decisions of 20th century: To tread on the Moon and fight versus cancer.

H42.   Two strategic decisions of 21st century: Human Genome Project and PPM.

Q10.   What prognosis do I have?

Personalized Medicine for Cancer

Q11. Which treatment do you recommend to me?

Q12. Which survival will I have?

H43. Changes in the oncologic paradigm: Medicine = all → PPM.

Q13. How to achieve sustainability?

A13. To prioritize value (maximize benefit) and contain costs.

Q14. Molecular determinations in care: Who does it pay?

Q15. Patient-1 → dead, patient-2 → alive, why?

Q16. Cancer classified into multiple molecular subtypes: How do people analyze them?

H44. (Blumenthal, 2017). IT, Targeted and Conventional Therapy in Metastatic NSCLC Trials .[110]

Q17. Are people able to look for a needle in a haystack?

Q18. (Lovly, 2016). LB: Why?

A18. Monitorize tumor charge; early detection; mutational-state detection; treatment monitorization; better representation.

H45. (Jaffee, 2017). Tumor Mutational Burden and Response Rate to PD-1 Inhibition.[111]

H46. PO: Cancer treatment model where therapeutic decisions are guided by each patient's molecular attributes.

H47. (Bonastre, 2016). The cost of molecular-guided therapy in oncology: A cost study.[112]

Q19. Doctor, tell me… what prognosis do I have?

A19. I do not know.

Q20. What treatment do you recommend me?

A20. I do not know.

Q21. What survival will I have?

A21. I do not know.

Q22. What quality of life?

A22. I do not know.

Q23. Will I be a long survivor?

A23. I do not know.

Q24. Molecular determinations in care: Who does he pay?

He provided the following conclusions on genomic studies (GSs).

230 Green Chemistry and Biodiversity: Principles, Techniques, and Correlations

C4. GSs allow detecting treatment directed after each therapeutic failure optimizing options.

C5. GSs in LB allows detecting mutations in the tumor when no available solid biopsy exists.

He provided the additional questions and answer.

Q25. (Molteni, 2018). Cancer diagnosis from a blood draw?

Q26. What have other countries done?

A26. Precision Medicine Initiative (US, 2015).

Q27. Is it possible to do it?

Q28. How do people do it?

Q29. What does it join all them?

Q30. Why a strategy of PPM?

He provided the following conclusions on why a strategy of PPM.

C6. To improve the clinical results in the patients betters in quality and quantity of life.

C7. Adapt people to scientific advances accelerating/facilitating access and needs.

C8. To contribute to the equity of access to precision care of maximum quality.

C9. Rationalization of health cost and national-health sustainability.

C10. Place Spain in PPM-strategies van.

H48. The traditional model is exhausted and does not serve people.

14.5 SUPERJ RESCUE! COMIC-STYLE RESEARCH: WHY IS THERE NO *CURE* VERSUS CANCER?

Sánchez Ruiz raised a Q on why there is no cure yet against *cancer* (*cf.* Fig. 14.3).[113]

Q1. Why is there no cure yet against cancer?

Personalized Medicine for Cancer

FIGURE 14.3 SuperJ to the rescue! Why is there no cure against cancer?
Source: Ref. [113].

14.6 DISCUSSION

The PPM means a change of paradigm. Intrinsically, every tumor–host complex is a rare disease. The BC is not a unique disease but a group of tumor subtypes with different natural history. The traditional model is exhausted and does not serve people.

Why is there no cure yet against cancer? Cancer is not a unique disease. Cancer cannot be treated as an infectious disease. To design therapies that destroy viruses or bacteria without affecting people's own cells results simpler. The problem is that tumor cells are cells of people's own body.

The PO increases the rates of survival. The PO allows safer treatments directed versus targets (decreases AEs). The PO removes the use of unnecessary or ineffective treatments. The PO improves system efficiency. Critical moment in PPM: Information is more important than drugs. Patients and healthcares search for results. Academics must lead to change. Advantage: They dispose of clinical and scientific knowledge.

14.7 FINAL REMARKS

From the present results, discussion and provisional conclusions, the final remarks can be drawn.

1. Precision oncology increases the rates of survival.
2. Precision oncology allows safer treatments directed versus targets (decreases adverse effects).
3. Precision oncology removes the use of unnecessary or ineffective treatments.
4. Precision oncology improves system efficiency.
5. A critical moment in precision medicine: Information is more important than drugs.
6. Patients and healthcares search for results.
7. Academics must lead change. Advantage: They dispose of clinical and scientific knowledge.
8. Luminal disease ER⁺ presents clinical benefit to therapies directed versus proper receptor.
9. Improvement from young women–hormone therapies duration in menopausal women.
10. Genetic platforms have prognostic value in patients with luminal breast cancer.
11. Genetic platforms are tool useful in decisions on adjuvant therapy in negative ganglia.
12. Controversy on GPs in decisions on adjuvant therapy in positive ganglia.
13. Prospective clinical studies that will provide information on points with controversy.
14. Anti*HER2* treatment changed the natural history of breast cancer.
15. Anti*HER2* adjuvant therapy plus chemotherapy reduces 50% relapses and deaths.

Personalized Medicine for Cancer

16. The optimal duration of therapy with Trastuzumab is one year.
17. Precocious tumors cannot be treated with not many toxic guides.
18. Pertuzumab adds benefit in tumors with axillary affectation and hormone receptor⁻.
19. The era of treating cancer in an equal manner for all has finished.
20. Understanding cancer molecular bases allows: Targets identification, therapeutic focuses.
21. Biomarkers identification made possible patients selection for different treatments.
22. The present challenge in oncology is to individualize treatment for every patient and tumor.
23. Gastric cancer is a heterogeneous disease.
24. New personalized therapies should be based on the molecular classification of every tumor.
25. Immunotherapy effective in microsatellite instability/Epstein–Barr virus gastric cancers.
26. Genomic studies in blood detect treatment directed after each therapeutic failure.
27. Blood genomic studies allow detecting mutations in the tumor without an available biopsy.
28. To improve the clinical results in the patients betters in quality and quantity of life.
29. Adapt people to scientific advances accelerating/facilitating access and needs.
30. To contribute to the equity of access to precision care of maximum quality.
31. Rationalization of health cost and national-health sustainability.
32. Place Spain in the van of precision-medicine strategies.

ACKNOWLEDGMENTS

The authors thank the support from Generalitat Valenciana (Project No. PROMETEO/2016/094) and Universidad Católica de Valencia *San Vicente Mártir* (Projects Nos. UCV.PRO.17-18.AIV.03 and 2019-217-001).

KEYWORDS

- **association**
- **breast cancer**
- **causation**
- **cancer genome**
- **cancer hypothesis**
- **computational systems biology**
- **genomic complexity**

REFERENCES

1. Torrens, F; Castellano, G. Modelling of Complex Multicellular Systems: Tumour–immune Cells Competition. *Chem. Central J.* **2009,** *3* (Suppl. I), 75–1–1.
2. Torrens, F.; Castellano, G. Information Theoretic Entropy for Molecular Classification: Oxadiazolamines as Potential Therapeutic Agents. *Curr. Comput.-Aided Drug Des.* **2013,** *9*, 241–253.
3. Torrens, F.; Castellano, G. Molecular Classification of 5Amino2aroylquinolines and 4Aroyl6,7,8trimethoxyquinolines as Highly Potent Tubulin Polymerization Inhibitors. *Int. J. Chemoinf. Chem. Eng.* **2013,** *3* (2), 1–26.
4. Estrela, J. M.; Mena, S.; Obrador, E.; Benlloch, M.; Castellano, G.; Salvador, R.; Dellinger, R. W. Polyphenolic Phytochemicals in Cancer Prevention and Therapy: Bioavailability versus bioefficacy. *J. Med. Chem.* **2017,** *60*, 9413–9436.
5. Torrens, F.; Castellano, G. *Molecular Classification of Antitubulin Agents with Indole Ring Binding at Colchicine-Binding Site.* In *Molecular Insight of Drug Design*; Parikesit, A. A., Ed.; InTechOpen: Vienna, 2018; pp 47–67.
6. Torrens, F.; Castellano, G. *Molecular Classification of 2Phenylindole3carbaldehydes as Potential Antimitotic Agents in Human Breast Cancer Cells.* In *Theoretical Models and Experimental Approaches in Physical Chemistry: Research Methodology and Practical Methods*; Haghi, A. K., Thomas, S., Praveen, K. M., Pai, A. R., Eds.; Apple Academic–CRC: Waretown, NJ, in press.
7. Torrens, F.; Castellano, G. *Cancer and Hypotheses on Cancer.* In *Molecular Chemistry and Biomolecular Engineering: Integrating Theory and Research with Practice*; Pogliani, L., Torrens, F., Haghi, A.K., Eds.; Apple Academic–CRC: Waretown, NJ, in press.
8. Torrens, F.; Castellano, G. AIDS Destroys Immune Defences: Hypothesis. *New Front. Chem.* **2014,** *23*, 11–20.
9. Torrens-Zaragozá, F.; Castellano-Estornell, G. Emergence, Spread, and Uncontrolled Ebola Outbreak. *Basic Clin. Pharmacol. Toxicol.* **2015,** 117 (Suppl. 2), 38–38.

Personalized Medicine for Cancer

10. Torrens, F.; Castellano, G. 2014 spread/uncontrolled Ebola outbreak. *New Front. Chem.* **2015**, *24*, 81–91.
11. Torrens, F.; Castellano, G. Ebola virus disease: Questions, Ideas, Hypotheses and Models. *Pharmaceuticals* **2016**, 9, 14–6-6.
12. Lluch, A., Ed. Book of Abstracts, Medicina Personalizada de Precisión: De la Teoría a la Práctica, València, Spain, September 17-19, 2018; Fundación Instituto Roche: València, Spain, 2018.
13. Pérez-Alonso, M. Book of Abstracts, Medicina Personalizada de Precisión: De la Teoría a la Práctica, València, Spain, September 17-19, 2018; Fundación Instituto Roche: València, Spain, 2018; O-1.
14. Tolosa, A. ¿Cuántos genes tiene el genoma humano? *Genética Médica* **2018**, *2018* (Jul. 9), 1–1.
15. Feero, W.G.; Guttmacher, A. E.; Collins, F. S. Genomic Medicine — An Updated Primer. *N. Engl. J. Med.* **2010**, *362*, 2001–2011.
16. Hamburg, M. A.; Collins, F. S. The Path to Personalized Medicine. *N. Engl. J. Med.* **2010**, *363*, 301–304.
17. Wang, L.; McLeod, H. L.; Weinshilboum, R. M. Genomics and Drug Response. *N. Engl. J. Med.* **2011**, 364, 1144–1153.
18. Goldberger, J. J.; Buxton, A. E. Personalized medicine vs Guideline-Based Medicine. *J. Am. Med. Assoc.* **2013**, *309*, 2559–2560.
19. McPherson, J. D. A Defining Decade in DNA Sequencing. *Nat. Methods* **2014**, 11, 1003–1005.
20. Biesecker, L. G.; Green, R. C. Diagnostic Clinical Genome and Exome Sequencing. *N. Engl. J. Med.* **2014**, *370*, 2418–2425.
21. Xue, Y.; Ankala, A.; Wilcox, W. R.; Hegde, M. R. Solving the Molecular Diagnostic Testing Conundrum for Mendelian Disorders in the Era of Next-generation Sequencing: Single-gene, Gene Panel, or Exome/Genome Sequencing. *Genet. Med.* **2014**, *17*, 444–451.
22. Yang, Y.; Muzny, D. M.; Reid, J. G.; Bainbridge, M. N.; Willis, A.; Ward, P. A.; Braxton, A.; Beuten, J.; Xia, F.; Niu, Z.; Hardison, M.; Person, R.; Bekheirnia, M. R.; Leduc, M. S.; Kirby, A.; Pham, P.; Scull, J.; Wang, M.; Ding, Y.; Plon, S. E.; Lupski, J. R.; Beaudet, A. L.; Gibbs, R. A.; Eng, C. M. Clinical Whole-exome Sequencing for the Diagnosis of Mendelian Disorders. *N. Engl. J. Med.* **2013**, *369*, 1502–1511.
23. Yang, Y.; Muzny, D.M.; Xia, F.; Niu, Z.; Person, R.; Ding, Y.; Ward, P.; Braxton, A.; Wang, M.; Buhay, C.; Veeraraghavan, N.; Hawes, A.; Chiang, T.; Leduc, M.; Beuten, J.; Zhang, J.; He, W.; Scull, J.; Willis, A.; Landsverk, M.; Craigen, W. J.; Bekheirnia, M. R.; Stray-Pedersen, A.; Liu, P.; Wen, S.; Alcaraz, W.; Cui, H.; Walkievicz, M.; Reid, J.; Bainbridge, M.; Patel, A.; Boerwinkle, E.; Beaudet, A. L.; Lupski, J. R.; Plon, S. E.; Gibbs, R. A.; Eng, C. M. Molecular Findings Among Patients Referred for Clinical Whole-Exome Sequencing. *J. Am. Med. Assoc.* **2014**, *312*, 1870–1879.
24. Topol, E. T. Individualized Medicine from Prewomb to Tomb. *Cell* **2014**, *157*, 241–253.
25. Kalokairinou, L.; Howard, H. C.; Borry, P. Changes on the Horizon for Consumer Genomics in the EU: Test Results May no Longer be Available Directly to Consumers. *Science* **2014**, *346*, 296–298.

236   Green Chemistry and Biodiversity: Principles, Techniques, and Correlations

26. Rahman, N. Realizing the Promise of Cancer Predisposition Genes. *Nature (London)* **2014**, *505*, 302–308.
27. Rupaimoole, R.; Slack, F. J. MicroRNA Therapeutics: Towards a New Era for the Management of Cancer and Other Diseases. *Nat. Rev. Drug Discov.* **2017**, *16*, 203–222.
28. Cerro-Herreros, E.; Sabater-Arcis, M.; Fernandez-Costa, J. M.; Moreno, N.; Perez-Alonso, M.; Llamusi, B.; Artero, R. *miR23b* and *miR218* Silencing Increase Muscleblind-like Expression and Alleviate Myotonic Dystrophy Phenotypes in Mammalian Models. *Nat. Commun.* **2018**, *9* 2482-1-13.
29. Tolosa, A. Disección del gen *BRCA1* para mejorar el diagnóstico genético en cáncer de mama y ovario. *Genética Médica* **2018**, *2018* (Sep. 14), 1–1.
30. Gallardo, M. Desarrollada una nueva metodología molecular para la detección de niveles mínimos de cálulas tumorales en leukemia. *Genética Médica* **2018**, *2018* (Aug. 22) 1–1.
31. Ribas, G. Book of Abstracts, Medicina Personalizada de Precisión: De la Teoría a la Práctica, València, Spain, September 17-19, 2018; Fundación Instituto Roche: València, Spain, 2018; O-2.
32. Jorde, L.; Carey, J.; Bamshad, M. *Medical Genetics*; Elsevier: Amsterdam, The Netherlands, 2015.
33. Kluger, J.; Park, A. The Angelina effect. *Time* **2013**, *2013* (May 27), 1–1
34. Jantús-Lewintre, E. Book of Abstracts, Medicina Personalizada de Precisión: De la Teoría a la Práctica, València, Spain, September 17-19, 2018; Fundación Instituto Roche: València, Spain, 2018; O-3.
35. Biomarker Definitions Working Group. Biomarkers and Surrogate Endpoints: Preferred Definitions and Conceptual Framework. *Clin. Pharmacol. Ther.* **2001**, *69*, 89–95.
36. Ashworth, A.; Hudson, T. J. Comparisons Across Cancers. *Nature (London)* **2013**, *502*, 306–307.
37. Chin, L.; Gray, J. W. Translating Insights from the Cancer Genome into Clinical Practice. *Nature (London)* **2008**, *452*, 553–563.
38. Calabuig-Fariñas, S.; Jantús-Lewintre, E.; Herreros-Pomares, A.; Camps. C. Circulating Tumor Cells Versus Circulating Tumor DNA in Lung Cancer — Which One will Win? *Transl. Lung Cancer Res.* **2016**, *5*, 466–482.
39. Arenas, J.; Camps, C.; Paramio, J.; Nieto, P., Eds. *Comprometidos con la Investigación en Cáncer: Primer Informe sobre la Investigación e Innovación en Cáncer en España*; Asociación Española contra el Cáncer–Asociación Española de Investigación sobre el Cáncer–Fundación Bancaria *La Caixa*: Madrid, Spain, 2018.
40. Llombart-Cussac, A. Book of Abstracts, Medicina Personalizada de Precisión: De la Teoría a la Práctica, València, Spain, September 17-19, 2018; Fundación Instituto Roche: València, Spain, 2018; O-4.
41. Ledermann, J.; Harter, P.; Gourley, C.; Friedlander, M.; Vergote, I.; Rustin, G.; Scott, C. L.; Meier, W.; Shapira-Frommer, R.; Safra, T.; Matei, D.; Fielding, A.; Spencer, S.; Daugherty, B.; Orr, M.; Hodgson, D.; Barrett, J.C.; Matulonis, U. Olaparib Maintenance Therapy in Patients with Platinum-sensitive Relapsed Serous OC: A Preplanned Retrospective Analysis of Outcomes by *BRCA* Status in a Randomised Phase 2 Trial. *Lancet Oncol.* **2014**, *15*, 852–861.

Personalized Medicine for Cancer 237

42. Cervantes, A. Book of Abstracts, Medicina Personalizada de Precisión: De la Teoría a la Práctica, València, Spain, September 17-19, 2018; Fundación Instituto Roche: València, Spain, 2018; O-5.
43. Dienstmann, R.; Tabernero, J. Cancer: A Precision Approach to Tumour Treatment. *Nature (London)* **2017**, *548*, 40–41.
44. Picco, G.; Garnett, M.J. A Road Map for Precision Cancer Medicine Using Personalized Models. *Cancer Discov.* **2017**, *7*, 456–458.
45. Doebele, R.C.; Davis, L. E.; Vaishnavi, A.; Le, A. T.; Estrada-Bernal, A.; Keysar, S.; Jimeno, A.; Varella-Garcia, M.; Aisner, D. L.; Li, Y.; Stephens, P. J.; Morosini, D.; Tuch, B. B.; Fernandes, M.; Nanda, N.; Low, J. A. An Oncogenic *NTRK* Fusion in a Patient with Soft-tissue Sarcoma with Response to the Tropomyosin-related Kinase Inhibitor LOXO101. *Cancer Discov.* **2015**, *5*, 1049–1055.
46. Drilon, A.; Laetsch, T. W.; Kummar, S.; DuBois, S. G.; Lassen, U. N.; Demetri, G. D.; Nathenson, M.; Doebele, R. C.; Farago, A. F.; Pappo, A. S.; Turpin, B.; Dowlati, A.; Brose, M. S.; Mascarenhas, L.; Federman, N.; Berlin, J.; El-Deiry, W.S.; Baik, C.; Deeken, J.; Boni, V.; Nagasubramanian, R.; Taylor, M.; Rudzinski, E. R.; Meric-Bernstam, F.; Sohal, D. P. S.; Ma, P. C.; Raez, L. E.; Hetchman, J. F.; Benayed, R.; Ladanyi, M.; Tuch, B. B.; Ebata, K.; Cruickshank, S.; Ku, N. C.; Cox, M. C.; Hawkins, D. S.; Hong, D. S.; Hyman, D. M. Efficacy of Larotrectinib in *TRK* Fusion-positive Cancers in Adults and Children. *N. Engl. J. Med.* **2018**, *378*, 731–739.
47. André, F. Developing Anticancer Drugs in Orphan Molecular Entities — A Paradigm Under Construction. *N. Engl. J. Med.* **2018**, *378*, 763–765.
48. Manssard, C.; Michiels, S.; Ferté, C.; Le Deley, M. C.; Lacroix, L.; Hollebecque, A.; Verlingue, L.; Ileana, E.; Rosellini, S.; Ammari, S.; Ngo-Camus, M.; Bahleda, R.; Gazzah, A.; Varga, A.; Postel-Vinay, S.; Loriot, Y.; Even, C.; Breuskin, I.; Auger, N.; Job, B.; De Beare, T.; Deschamps, F.; Vielh, P.; Scoazec, J. Y.; Lazar, V.; Richon, C.; Ribrag, V.; Deutsch, E.; Angevin, E.; Vassal, G.; Eggermont, A.; André, F.; Soria, J. C. High-throughput Genomics and Clinical Outcome in Hard-to-treat Advanced Cancers: Results of the MOSCATO 01 trial. *Cancer Discov.* **2017**, *7*, 586–595.
49. Kavuri, S.M.; Jain, N.; Galimi, F.; Cottino, F.; Leto, S.M.; Migliardi, G.; Searleman, A.C.; Shen, W.; Monsey, J.; Trusolino, L.; Jacobs, S.A.; Bertotti, A.; Bose, R. *HER2* Activating Mutations are Targets for Colorectal Cancer Treatment. *Cancer Discov.* 2015, *5*, 832–841.
50. Sartore-Bianchi, A.; Trusolino, L.; Martino, C.; Bencardino, K.; Lonardi, S.; Bergamo, F.; Zagonel, V.; Leone, F.; Depetris, I.; Martinelli, E.; Troiani, T.; Ciardello, F.; Racca, P.; Bertotti, A.; Siravegna, G.; Torri, V.; Amatu, A.; Ghezzi, S.; Marrapese, G.; Palmeri, L.; Valtorta, E.; Cassigena, A.; Lauricella, C.; Vanzulli, A.; Regge, D.; Veronese, S.; Comoglio, P. M.; Bardelli, A.; Marsoni, S.; Siena, S. Dual-targeted therapy with Trastuzumab and Lapatinib in Treatment-refractory, *HRAS* codon 12/13 wild-type, *HER2*Positive Metastatic Colorectal Cancer (HERACLES): A Proof-of-Concept, Multicentre, Open-label, Phase 2 Trial. *Lancet Oncol.* **2016**, *17*, 738–746.
51. Pauli, C.; Hopkins, B. D.; Prandi, D.; Shaw, R.; Fedrizzi, T.; Sboner, A.; Sailer, V.; Augello, M.; Puca, L.; Rosati, R.; McNary, T. J.; Churakova, Y.; Cheung, C.; Triscott, J.; Pisapia, D.; Rao, R.; Mosquera, J. M.; Robinson, B.; Faltas, B. M.; Emerling, B. E.; Gadi, V. K.; Bernard, B.; Elemento, O.; Beltran, H.; Demichelis, F.; Kemp, C. J.;

Grandori, C.; Cantley, L. C.; Rubin, M. A. Personalized In Vitro and In Vivo Cancer Models to Guide Precision Medicine. *Cancer Discov.* **2017,** *7,* 462–477.

52. Vlachogiannis, G.; Hedayat, S.; Vatsiou, A.; Jamin, Y.; Fernández-Mateos, J.; Khan, K.; Lampis, A.; Eason, K.; Huntingford, I.; Burke, R.; Rata, M.; Koh, D. M.; Tunariu, N.; Collins, D.; Hulkki-Wilson, S.; Ragulan, C.; Spiteri, I.; Moorcraft, S. Y.; Chau, I.; Rao, S.; Watkins, D.; Fotiadis, N.; Bali, M.; Darvish-Damavandi, M.; Lote, H.; Eltahir, Z.; Smyth, E. C.; Begum, R.; Clarke, P. A.; Hahne, J. C.; Dowsett, M.; de Bono, J.; Workman, P.; Sadanandam, A.; Fassan, M.; Sansom, O. J.; Eccles, S.; Starling, N.; Braconi, C.; Sottoriva, A.; Robinson, S. P.; Cunningham, D.; Valeri, N. Patient-derived Organoids Model Treatment Response of Metastatic Gastrointestinal cancers. *Science* **2018,** *359,* 920–926.

53. Lluch, A. Book of Abstracts, Medicina Personalizada de Precisión: De la Teoría a la Práctica, València, Spain, September 17-19, 2018; Fundación Instituto Roche: València, Spain, 2018; O-6.

54. Rodriguez, M. A.; Walters, R. S.; Burke, T. W., Eds. *60 Years of Survival Outcomes at the University of Texas MD Anderson Cancer Center*; Springer: New York, NY, 2013.

55. Walgren, R. A.; Meucci, M. A.; McLeod, H. L. Pharmacogenomic Discovery Approaches: Will the Real Genes Please Stand Up? NotPersonalized Treatment. *J. Clin. Oncol.* **2005,** *23,* 7342–7349.

56. Hanahan, D.; Weinberg, R. A. The Hallmarks of Cancer. *Cell* **2000,** *100,* 57–70.

57. Weigelt, B.; Reis-Filho, J. S. Histological and Molecular Types of Breast Cancer: Is there a Unifying Taxonomy? *Nat. Rev. Clin. Oncol.* **2009,** *6,* 718–730.

58. Early Breast Cancer Trialists' Collaborative Group (EBCTCG), Davies, C.; Godwin, J.; Gray, R.; Clarke, M.; Cutter, D.; Darby, S.; McGale, P.; Pan, H. C.; Taylor, C.; Wang, Y. C.; Dowsett, M.; Ingle, J.; Peto, R. Relevance of Breast Cancer Hormone Receptors and Other Factors to the Efficacy of Adjuvant Tamoxifen: Patient-level Metaanalysis of Randomised Trials. *Lancet* **2011,** *378,* 771–784.

59. Rodon, J.; Dienstmann, R.; Serra, V.; Tabernero, J. Development of PI3K Inhibitors: Lessons Learned from Early Clinical Trials. *Nat. Rev. Clin. Oncol.* **2013,** *10,* 143–153.

60. Fruman, D. A.; Rommel, C. PI3K and cancer: Lessons, Challenges and Opportunities. *Nat. Rev. Drug Discov.* **2014,** *13,* 140–156.

61. Foulkes, W. D.; Smith, I. E.; Reis-Filho, J.S. Current Concepts: Triple-Negative Breast Cancer. *N. Engl. J. Med.* **2010,** *363,* 1938–1948.

62. Aarts, M.; Linardopoulos, S.; Turner, N.C. Tumour Selective Targeting of Cell Cycle Kinases for Cancer Treatment. *Curr. Opin. Pharmacol.* **2013,** *13,* 529–535.

63. Kassam, F.; Enright, K.; Dent, R.; Dranitsaris, G.; Myers, J.; Flynn, C.; Fralick, M.; Kumar, R.; Clemons, M. Survival Outcomes for Patients with Metastatic Triple-negative Breast Cancer: Implications for Clinical Practice and Trial Design. *Clin. Breast Cancer* **2009,** *9,* 29–33.

64. Dent, R.; Trudeau, M.; Pritchard, K. I.; Hana, W. M.; Kahn, H. K.; Sawka, C. A.; Lickley, L. A.; Rawlinson, E.; Sun, P.; Narod, S. A. Triple-negative Breast Cancer: Clinical Features and Patterns of Recurrence. *Clinical Cancer Res.* **2007,** *13,* 4429–4434.

65. Yarden, Y.; Sliwkowski, M. X. Untangling the ErbB Signalling Network. *Nat. Rev. Mol. Cell Biol.* **2001,** *2,* 127–137.

Personalized Medicine for Cancer

66. DeVita, V. T., Jr.; Lawrence, T. S.; Rosenberg, S. A. *DeVita, Hellman, and Rosenberg's Cancer: Principles & Practice of Oncology*; LWW: New York, NY, 2014.

67. Von Minckwitz, G.; Procter, M.; de Azambuja, E.; Zardavas, D.; Benyunes, M.; Viale, G.; Suter, T.; Arahmani, A.; Rouchet, N.; Clark, E.; Knott, A.; Lang, I.; Levy, C.; Yardley, D. A.; Bines, J.; Gelber, R. D.; Piccart, M.; Baselga, J. Adjuvant Pertuzumab and Trastuzumab in Early *HER2*Positive Breast Cancer. *N. Engl. J. Med.* **2017**, *377*, 122–131.

68. Gianni, L.; Pienkowski, T.; Im, Y. H.; Roman, L.; Tseng, L. M.; Liu, M. C.; Lluch, A.; Staroslawska, E., de la Haba-Rodriguez, J.; Im, S. A.; Pedrini, J. L.; Poirier, B.; Morandi, P;. Semiglazov, V.; Srimuninnimit, V.; Bianchi, G.; Szado, T.; Ratnayake, J.; Ross, G.; Valagussa, P. Efficacy and Safety of Neoadjuvant Pertuzumab and Trastuzumab in Women with Locally Advanced, Inflammatory, or Early HER2-positive Breast Cancer (NeoSphere): A Randomised Multicentre, Open-label, Phase 2 Trial. *Lancet Oncol.* **2012**, *13*, 25–32.

69. Berrocal, A. Book of Abstracts, Medicina Personalizada de Precisión: De la Teoría a la Práctica, València, Spain, September 17-19, 2018, Fundación Instituto Roche: València, Spain, 2018; O-7.

70. Dranoff, G. Cytokines in Cancer Pathogenesis and Cancer Therapy. *Nat. Rev. Cancer* **2004**, *4*, 11–22.

71. Male, D.; Brostoff, J.; Roth, D.; Roitt, I., Eds. *Immunology*; Elsevier: Amsterdam, The Netherlands, 2012.

72. Kahan, B.D. Individuality: The Barrier to Optimal Immunosuppression. *Nat. Rev. Immunol.* **2003**, *3*, 831–838.

73. Chen, L.; Flies, D.B. Molecular Mechanisms of T Cell Costimulation and Coinhibition. *Nat. Rev. Immunol.* **2013**, *13*, 227–242.

74. Coomes, S. M.; Pelly, V. S.; Wilson, M. S. Plasticity Within the $\alpha\beta^+CD4^+$ T-cell Lineage: When, How and What for? *Open Biol.* **2013**, *3*, 120157-1-15.

75. Dunn, G.P.; Old L.J.; Schreiber, R.D. The Immunology of Cancer Immunosurveillance and Immunoediting. *Immunity* **2004**, *21*, 137–148.

76. Vesely, M.D.; Kershaw, M.H.; Schreiber, R.D.; Smyth, M.J. Natural Innate and Adaptive Immunity to Cancer. *Ann. Rev. Immunol.* **2011**, *29*, 235–271.

77. D. Hanahan, D.; Weinberg, R.A. Hallmarks of Cancer: The Next Generation. *Cell* **2011**, *144*, 646–674.

78. Wolchok, J.D.; Hoos, A.; O'Day, S.; Weber, J.S.; Hamid, O.; Lebbé, C.; Maio, M.; Binder, M.; Bohnsack, O.; Nichol, G.; Humphrey, R.; Hodi, F.S. Guidelines for the Evaluation of Immune Therapy Activity in Solid Tumors: Immune-Related Response Criteria. *Clin. Cancer Res.* **2009**, *15*, 7412–7420.

79. Espinosa, E.; Márquez-Rodas, I.; Soria, A.; Berrocal, A.; Manzano, J. L.; Gonzalez-Cao, M.; Martin-Algarra, S. Predictive Factors of Response to Immunotherapy — A Review from the Spanish Melanoma Group (GEM). *Ann. Transl. Med.* **2017**, *5*, 389-1-7.

80. Aguiar, P.N., Jr.; De Mello, R.A.; Hall, P.; Tadokoro, H.; Lima Lopes, G. PD-L1 Expression as a Predictive Biomarker in Advanced Nonsmall-cell Lung Cancer: Updated Survival Data. *Immunotherapy* **2017**, *9*, 499–506.

81. Snyder, A.; Makarov, V.; Merghoub, T.; Yuan, J.; Zaretsky, J.M.; Desrichard, A.; Walsh, L. A.; Postow, M. A.; Wong, P.; Ho, T. S.; Hollmann, T. J.; Bruggeman, C.;

Kannan, K.; Li, Y.; Elipenahli, C.; Liu, C.; Harbison, C. T.; Wang, L.; Ribas, A.; Wolchok, J. D.; Chan, T. A. Genetic Basis for Clinical Response to CTLA-4 Blockade in Melanoma. *N. Engl. J. Med.* **2014**, *371*, 2189–2199.

82. Hellmann, M. D.; Nathanson, T.; Rizvi, H.; Creelan, B. C.; Sanchez-Vega, F.; Ahuja, A.; Ni, A.; Novic, J. B.; Mangarin, L. M. B.; Abu-Akeel, M.; Liu, C.; Sauter, J. L.; Rekhtman, N.; Chang, E.; Callahan, M. K.; Chaft, J. E.; Voss, M. H.; Tenet, M.; Li, X. M.; Covello, K.; Renninger, A.; Vitazka, P.; Geese, W. J.; Borghael, H.; Rudin, C. M.; Antonia, S. J.; Swaton, C.; Hammerbacher, J.; Merghoub, T.; McGranahan, N.; Snyder, A.; Wolchok, J. D. Genetic Features of Response to Combination Immunotherapy in Patients with Advanced Nonsmall-cell Lung Cancer. *Cancer Cell* **2018**, *33*, 843–852.

83. Rosemberg, S.A.; Restifo, N.P.; Yang, C.J.; Morgan, R.A.; Dudley, M.E. Adoptive Cell Transfer: A Clinical Path to Effective Cancer Immunotherapy. *Nat. Rev. Cancer* **2008**, *8*, 299–308.

84. Rosemberg, S.A.; Restifo, N.P. Adoptive Cell Transfer as Personalized Immunotherapy for Human Cancer. *Science* **2015**, *348*, 62–68.

85. Ping, Y.; Liu, C.; Zhang, Y. T-cell Receptor-engineered T Cells for Cancer Treatment: Current Status and Future Directions. *Protein Cell* **2018**, *9*, 254–266.

86. Overwijk, W. W.; Wang, E.; Marincola, F. M.; Rammensee, H. G.; Restifo, N. P. Mining the Mutanome: Developing Highly Personalized Immunotherapies Based on Mutational Analysis of Tumors. *J. Immunother. Cancer* **2013**, *1*, 11–1-4.

87. Pérez-Fidalgo, J.A. Book of Abstracts, Medicina Personalizada de Precisión: De la Teoría a la Práctica, València, Spain, September 17-19, 2018; Fundación Instituto Roche: València, Spain, 2018; O-8.

88. The Cancer Genome Atlas Research Network. Integrated Genomic Analyses of Ovarian Carcinoma. *Nature (London)* **2011**, *474*, 609–615.

89. Morgan, T.H. *A Critique of the Theory of Evolution*; Princeton University: Princeton, NJ, 2016.

90. Lee, J. M.; Nair, J.; Zimmer, A.; Lipkowitz, S.; Annunziata, C. M.; Merino, M. J.; Swisher, E. M.; Harrell, M. I.; Trepel, J. B.; Lee, M. J.; Bagheri, M. H.; Botesteanu, D. A.; Steinberg, S. M.; Minasian. L.; Ekwede, I.; Kohn, E. C. Prexasertib, a cell cycle checkpoint kinase 1 and 2 inhibitor, in *BRCA* wild-type recurrent high-grade serous OC: A first-in-class proof-of-concept phase 2 study. *Lancet Oncol.* **2018**, *19*, 207–215.

91. S. Lheureux, S.; Oza, A.M. Endometrial Cancer — Targeted Therapies Myth or Reality? Review of Current Targeted Treatments. *Eur. J. Cancer* **2016**, *59*, 99–108.

92. Roda Pérez, D. Book of Abstracts, Medicina Personalizada de Precisión: De la Teoría a la Práctica, València, Spain, September 17-19, 2018; Fundación Instituto Roche: València, Spain, 2018; O-9.

93. The Cancer Genome Atlas Research Network. Comprehensive Molecular Characterization of Gastric Adenocarcinoma. *Nature (London)* **2014**, *513*, 202–209.

94. Torre, L.A.; Bray, F.; Siegel, R.L.; Ferlay, J.; Lortet-Tieulent, J.; Jemal, A. Global Cancer Statistics, 2012. *CA: Cancer J. Clin.* **2015**, *65*, 87–108.

95. De Angelis, R.; Sant, M.; Coleman, M. P.; Francisci, S.; Baili, P.; Pierannunzio, D.; Trama, A.; Visser, O.; Brenner, H.; Ardanaz, E.; Bielska-Lasota, M.; Engholm, G.; Nennecke, A.; Siesling, S.; Berrino, F.; Capocaccia, R.; the EUROCARE-5 Working

Personalized Medicine for Cancer 241

Group. Cancer Survival in Europe 1999–2007 by Country and Age: Results of EUROCARE-5 — A Population-based Study. *Lancet Oncol.* **2014**, *15*, 23–34.

96. Fan, H.; Chu, J.Y. The MMR Pathway Repairs Base-base Mismatches and Insertion and Deletion Loops. *Genomics, Proteomics, Bioinformatics* **2007**, *5*, 7–14.

97. Lordick, F.; Kang, Y. K.; Chung, H. C.; Salman, P.; Oh, S. C.; Bodoky, G.; Kurteva, G.; Volovat, C.; Moiseyenka, V. M.; Gorbunova, V.; Park, J. O.: Sawaki, A.; Celik, I.; Götte, H.; Melezínková, H.; Moehler, M. Capecitabine and *cis*platin with or without Cetuximab for Patients with Previously Untreated Advanced Gastric Cancer (EXPAND): A Randomised, Open-lab Phase 3 Trial. *Lancet Oncol.* **2013**, *14*, 490–499.

98. Metzger-Filho, O.; Winer, E.P.; Krop, I. Pertuzumab: Optimizing *HER2* blockade. *Clin. Cancer Res.* **2013**, *19*, 5552–5556.

99. Bang, Y.J.; Van Cutsem, E.; Feyereislova, A.; Chung, H.C.; Shen, L.; Sawaki, A.; Landick, F.; Obtsu, A.; Omura, Y.; Satoh, T.; Aprile, G.; Kulikov, E.; Hill, J.; Lehle, M.; Rüschoff, J.; Kang, Y.K. Trastuzumab in Combination with Chemotherapy versus Chemotherapy Alone for Treatment of *HER2*-Positive Advanced Gastric or Gastro-oesophageal Junction Cancer (ToGA): A Phase 3, Open-label, Randomised Controlled Trial. *Lancet* **2010**, *376*, 687–697.

100. Lordick, F.; Janjigian, Y. Y. Clinical Impact of Tumour Biology in the Management of Gastroesophageal Cancer. *Nat. Rev. Clin. Oncol.* **2016**, *13*, 348–360.

101. Tabernero, J.; Hoff, P. M.; Shen, L.; Ohtsu, A.; Shah, M. A.; Cheng, K.; Song, C.; Wu, H.; Eng-Wong, J.; Kang, Y. K. Pertuzumab (P) + Trastuzumab (H) + Chemotherapy (CT) for *HER2*-Positive Metastatic Gastric or Gastrooesophageal Junction Cancer (mGC/GEJC): Final Analysis of a Phase III Study (JACOB). *Ann. Oncol.* **2017**, *28* (Supplement 5), v209–v268–616O–1–1.

102. Thuss-Patience, P. C.; Shah, M. A.; Ohtsu, A.; Van Cutsem, E.; Ajani, J. A.; Castro, H.; Mansoor, W.; Chung, H. C.; Bodoky, G.; Shitara, K.; Phillips, G. D. L.; van der Horst, T.; Harle-Yge, M. L.; Althaus, B. L.; Kang, Y. K. Trastuzumab emtansine versus Taxane use for Previously Treated *HER2*-positive Locally Advanced or Metastatic Gastric or Gastro-oesophageal Junction Adenocarcinoma (GATSBY): An International Randomised, Open-label, Adaptive, Phase 2/3 Study. *Lancet Oncol.* **2017**, *18*, 640–653.

103. Jung, Y.D.; Mansfield, P.F.; Akagi, M.; Takeda, A.; Liu, W.; Bucana, C.D.; Hiklin, D.J.; Ellis, D.M. Effects of Combination Antivascular Endothelial Growth Factor Receptor and Antiepidermal Growth Factor Receptor Therapies on the Growth of Gastric Cancer in a Nude Mouse Model. *Eur. J. Cancer* **2002**, *38*, 1133–1140.

104. Fuchs, C. S.; Tomasek, J.; Yong, C. J.; Dumitru, F.; Passalacqua, R.; Goswami, C.; Safran, H.; dos Santos, L. V.; Aprile, G.; Ferry, D. R.; Melichar, B.; Tehfe, M.; Topuzov, E.; Zalcberg, J. R.; Chau, I.; Campbell, W.; Sivanandan, C.; Pikiel, J.; Koshiji, M.; Hsu, Y.; Liepa, A. M.; Gao, L.; Schwartz, J. D.; Tabernero, J. Ramucirumab Monotherapy for Previously Treated Advanced Gastric or Gastrooesophageal Junction Adenocarcinoma (REGARD): An International, Randomised, Multicentre, Placebo-controlled, Phase 3 Trial. *Lancet* **2014**, *383*, 31–39.

105. Wilke, H.; Muro, K.; Van Cutsem, E.; Oh, S. C.; Bodoky, G.; Shimada, Y.; Hironaka, S.; Sugimoto, N.; Lipatov, O.; Kim, T. Y.; Cunningham, D.; Rougier, P.; Kamatsu, Y.; Ajani, J.; Emig, M.; Carlesi, R.; Ferry, D.; Chandrawansa, K.; Schwartz, J.

242 Green Chemistry and Biodiversity: Principles, Techniques, and Correlations

D.; Ohtsu, A. Ramucirumab plus Paclitaxel versus Placebo Plus Paclitaxel in Patients with Previously Treated Advanced Gastric or Gastrooesophageal Junction Adenocarcinoma (RAINBOW): A Double-blind, Randomised Phase 3 Trial. *Lancet Oncol.* **2014,** *15,* 1224–1235.

106. Muro, K.; Chung, H.C.; Shankaran, V.; Geva, R.; Catenacci, D.; Gupta, S.; Elder, J.P.; Golan, T.; Le, D. T.; Burtness, B.; McRee, A. J.; Lin, C. C.; Pathiraja, K.; Lunceford, J.; Emancipator, K.; Juco, J.; Koshiji, M.; Bang, Y. J. Pembrolizumab for Patients with PD-L1-Positive Advanced Gastric Cancer (KEYNOTE-012): A Multicentre, Open-label, Phase 1b Trial. *Lancet Oncol.* **2016,** *17,* 717–726.

107. Kim, S. T.; Critescu, R.; Bass, A. J.; Kim, K. M.; Odegaard, J. I.; Kim, K.; Liu, X. Q.; Sher, X.; Jung, H.; Lee, M.; Lee, S.; Park, S. H.; Park, J. O.; Park, Y. S.; Lim, H. Y.; Lee, H.; Choi, M.; Talasaz, A.; Kang, P. S.; Cheng, J.; Loboda, A.; Lee, J.; Kang, W. K. Comprehensive Molecular Characterization of Clinical Responses to PD-1 Inhibition in Metastatic Gastric Cancer. *Nat. Med.* **2018,** *24,* 1449–1458.

108. The Cancer Genome Atlas Research Network. Comprehensive Molecular Characterization of Gastric Adenocarcinoma. *Nature (London)* **2014,** 513, 202–209.

109. Camps, C. Book of Abstracts, Medicina Personalizada de Precisión: De la Teoría a la Práctica, València, Spain, September 17-19, 2018; Fundación Instituto Roche: València, Spain, 2018; O-10.

110. Blumenthal, G. M.; Zhang, L.; Zhang, H.; Kazandjian, D.; Khozin, S.; Tang, S.; Goldberg, K.; Sridhara, R.; Keegan, P.; Pazdur, R. Milestone Analyses of Immune Checkpoint Inhibitors, Targeted Therapy, and Conventional Therapy in Metastatic Nonsmall Cell Lung Cancer Trials: A Metaanalysis. *J. Am. Med. Assoc. Oncol.* **2017,** *3,* e171029–1-1.

111. Yarchoan, M.; Hopkins, A.; Jaffee, E.M. Tumor Mutational Burden and Response Rate to PD-1 Inhibition. *N. Engl. J. Med.* **2017,** *377,* 2500–2501.

112. Pagès, A.; Foulon, S.; Zou, Z.; Lacroix, L.; Lemare, F.; de Baère, T.; Massard, C.; Soria, J.C.; Bonastre, J. The Cost of Molecular-guided Therapy in Oncology: A Prospective Cost Study Alongside the MOSCATO Trial. *Gent. Med.* **2017,** *19,* 683–690.

113. Sánchez Ruiz, J. ¡SuperJ al recate! La investigación en clave de cómic: ¿Por qué no hay aún una *cura* contra el *cáncer*? *Cris T Cuenta... (Madrid)* **2017,** *2017* (8), 8–8.

CHAPTER 15

Design, Synthesis, and Studies of Novel Piperidine Substituted Triazine Derivatives as Potential Anti-Inflammatory and Antimicrobial Agents

RAVINDRA S. SHNIDE

Department of Chemistry and Industrial Chemistry, Dayanand Science College, Latur 413512, Maharashtra, India

*Corresponding author. E-mail: rshinde.33381@gmail.com

ABSTRACT

A series of some novel 2,4-dimethoxy-6-(piperazin-1-yl)-1,3,5-triazine aryl ureido/aryl amido derivatives of biological interest were prepared and screened for their pro-inflammatory cytokines (TNF-α and IL-6) and antimicrobial activity. Among all the series of compounds screened, many compounds were found to have promising anti-inflammatory activity (up to 65–73% TNF-α and 73–85% IL-6 inhibitory activity) at concentration of 10 μM with reference to standard dexamethasone (75% TNF-α and 84% IL-6 inhibitory activities at 1 μM) while the similar compounds found to be potent antimicrobial agent showing even 2–2.5-fold more potency than that of standard ciprofloxacin and miconazole at the same MIC value of 10 μg/mL.

15.1 INTRODUCTION

The heterocyclic compounds have unique importance in the field of medicinal chemistry. There are 12.6 million of medicinal drugs currently registered of which about 6–7 millions of them are heterocyclic. The Cyanuric chloride is an attractive molecule for dendrimer synthesis owing to its low cost and chemoselective reactivity.[1] The triazine scaffold provides the basis for design of biologically relevant molecule with widespread application as therapeutics.[2] Considering the significance of heterocyclic compounds and their diverse applications, it has become an interesting research go for synthesis of novel triazine linked piperazine scaffolds containing phenyl piperazine carboxiamide urea and methanone amide derivatives.

The piperazine is an interesting heterocyclic moiety as constituent of several biologically active molecules.[3] Ananda Kumar et al.[4] revealed that, the scaffold piperazine and its analogues are an important pharmacophore that can be found in biologically active compounds across a number of different therapeutic areas. The piperazine core has ability of binding to many receptors with elevated affinity and therefore piperazine has a functional structure. For instance, linezolid, eperezolid, AZD2563, and itraconazole, which are the antibiotics used in the treatment of microbial infections, contain a piperazine ring in their structures.[5] The nitrogen in the piperazine ring plays an important role in biological research and drug manufacturing industry.[6] Piperazinyl attached Ciprofloxacin dimers are efficient antibacterial,[7] antipsychotic, and antimalarial agents.[8,9] Further, diphenyl piperazine analogues acquire broad pharmacological action on central nervous system (CNS) and dopaminergic neurotransmission. They have included potent and selective DA uptake inhibitors and some have been in preclinical development or clinical trial.[10] Some piperazine quinoline combination is found in the structure of many well-known drugs like Ciprofloxacin, norfloxacin, ofloxacin, pefloxacin, rufloxacin, enrofloxacin, and so on. There are many reports, that is, Chikhalia et al.[11] on triazines linked with piperazine or aniline scaffolds acting as an excellent antimicrobial agent against various diseases caused by fungi and bacteria. The triazine-linked piperazines are also studied for their effect in cellular, biochemical, tissue, animal-based assays. The urea derivatives have been investigated for their biological activities including such as N-alkyl urea hydroxamic acids as a new class of PDF inhibitor. This class of compounds has potent whole cell activity against both gram positive and gram negative bacteria but is devoid of MMP inhibition.[12] During the last decade, several 1,3,5-triazine derivatives were synthesized as useful chemotherapeutic agents for the treatment of tumor diseases.[13] The N-2,4-pyrimidine-N,

Design, Synthesis, and Studies of Novel Piperidine 245

N-phenyl/alkyl urea acts as inhibitor of tumor necrosis factor alpha (TNF-α).[14,15] Some substituted urea derivatives were reported as a potent TNF-α production inhibitors. Result are based on literature survey[16] above, these novel three substituted urea derivatives may prove be potential inhibitor for treatment of diseases mediated by TNF-α.

Many N-substituted piperazine annulated s-triazine derivatives were screened for their antimicrobial activity against different strains. Hence, it is concluded that trisubstituted s-triazines bearing piperazine nucleus are proved to be beneficial in augmenting antimicrobial probe by Desai et al.[17]

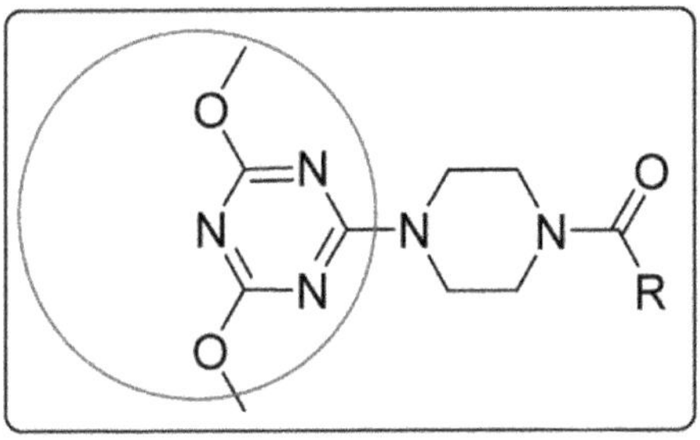

FIGURE 15.1 Parent core.

The literature survey reveals that the triazine analogues and piperazines are very important in medicinal chemistry and also in drug discovery process. On the basis of recent literature and our logical research to improve the anti-inflammatory and antimicrobial activity of the compound based on triazine core and related scaffold (Fig. 15.1) it is considered worth to investigate that hybrid compounds incorporating a piperazine core and urea/amide in single molecular frame could produce novel potent anti-inflammatory and antimicrobial agents. Hence, herein we report the synthesis, anti-inflammatory, and antimicrobial activity evaluation study of novel 4-(4,6-dimethoxy-1,3,5-triazin-2-yl)-N-substituted phenylpiperazine-1-carboxamide urea derivatives and (4-(4,6-dimethoxy-1,3,5-triazin-2-yl) piperazin-1-yl) (substituted phenyl) methanone amide derivatives.

15.2 PRESENT WORK

A series of novel 2,4-dimethoxy-6-(piperazin-1-yl)-1,3,5-triazine aryl ureido/aryl amido derivatives of biological interest have been designed and synthesized. The compounds 4-(4,6-dimethoxy-1,3,5-triazin-2-yl)-*N*-substituted phenyl piperazine-1-carboxamide urea derivatives (3a–h) were synthesized from 2-chloro-4,6-dimethoxy-1,3,5-triazine (3A), tert-butyl piperazine-1-carboxylate (3B) in presence of K_2CO_3 in DMF at 70–80°C to give corresponding tert-butyl 4-(4,6-dimethoxy-1,3,5-triazin-2-yl)piperazine-1-carboxylate (3C) which was deprotected to obtain 2, 4-dimethoxy-6-(piperazin-1-yl)-1,3,5-triazine (3D) and reacted with substituted aryl isocyanate to get desired products (3a–h). Similarly, (4-(4,6-dimethoxy-1,3,5-triazin-2-yl) piperazin-1-yl) (substituted phenyl) methanone amide derivatives (3i–p) were synthesized from 2-chloro-4,6-dimethoxy-1,3,5-triazine (3A), tert-butyl piperazine-1-carboxylate (3B) in presence of K_2CO_3 in DMF at 70–80°C to give corresponding tert-butyl 4-(4,6-dimethoxy-1,3,5-triazin-2-yl)piperazine-1-carboxylate (3C) which was deprotected to give intermediate (3D) and reacted with substituted benzoyl chloride to obtain the final products.

Our synthetic strategy for novel compounds was outlined in Scheme 15. 1. The 2-chloro-4,6-dimethoxy-1,3,5-triazine (3A) treated with tert-butyl piperazine-1-carboxylate (3B) under basic condition to give tert-butyl 4-(4,6-dimethoxy-1,3,5-triazin-2-yl)piperazine-1-carboxylate (3C). On acidic deprotection of 3 gives common intermediate (3D) which is treated with respective substituted aryl isocyanate and substituted benzoyl chloride to give 4-(4,6-dimethoxy-1,3,5-triazin-2-yl)-*N*-substituted phenylpiperazine-1-carboxamide urea derivatives (3a–h) and (4-(4,6-dimethoxy-1,3,5-triazin-2-yl)piperazin-1-yl)(substituted phenyl) methanone amide derivatives (3i–p), respectively.

All the new synthesized compounds were evaluated for their anti-inflammatory activity (TNF-α and IL-6 inhibitory activity) and antimicrobial (antifungal and antibacterial) activities against some selected pathogenic bacteria and fungi.

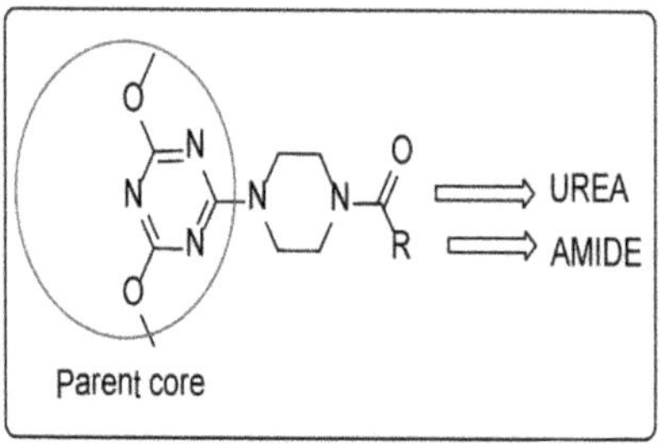

Parent core

FIGURE 15.2   Design scaffold.

Design, Synthesis, and Studies of Novel Piperidine

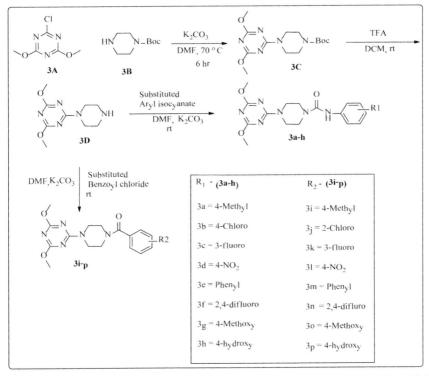

SCHEME 15.1 2, 4-dimethoxy-6-(piperazin-1-yl)-1,3,5-triazine aryl substituted ureido/aryl amido derivatives.

15.3 EXPERIMENTAL

15.3.1 MATERIALS AND METHODS

All chemicals and solvents used are of reagents grade and were used without further purification. The IR spectra were recorded on FT-IR Bruker with KBr disc and the ^{1}H NMR spectra were recorded with Bruker 400 MHz AVANCE instrument and J values are in Hertz and chemical shifts (δ) are reported in ppm relative to internal reference tetramethylsilane. The mass spectra were recorded with Thermo Finnigan-TSQ Quarter Ultra (triple Quad) at S.A.I.F. Division, Punjab University, Chandigarh.

248 Green Chemistry and Biodiversity: Principles, Techniques, and Correlations

15.3.1.1 SYNTHESIS OF TERT-BUTYL 4-(4,6-DIMETHOXY-1,3,5-TRIAZIN-2-YL) PIPERAZINE-1-CARBOXYLATE (3C)

3A   3B   3C

15.3.1.1.1 Procedure

The mixture of 2-chloro-4,6-dimethoxy-1,3,5-triazine (3A) (0.01 mole, 1.75 g), tert-butyl piperazine-1-carboxylate (3B) (0.01 mole, 3.25 g) and K_2CO_3 (0.02 mole, 2.75 g) in DMF (for 1 g, 25 mL) was heated at 70–80°C for 6 h. The progress of this reaction was neatly monitored by TLC (5% MeOH/DCM). After cooling the reaction mixture, water was added to give white precipitate which on filtration gives off a white solid as title compound (3C). The solid compound obtained was dried under vacuum.

15.3.1.1.2 Physicochemical Data

Yield: 80% Mol. Weight: 325.12   Mol. Formula: $C_{14}H_{23}N_5O_4$

^1H NMR (DMSO-d6, 400 MHz): δ 3.72 (s, 6H, $-2OCH_3$), 3.55 (m, 4H, $-CH_2-CH_2-$), 3.02 (m, 4H, $-CH_2-CH_2-$), 1.25 (s, 9H, $-C(-CH_3)_3$).
MS (APCI) m/z: 326[M+1]$^+$. Anal. Calcd for $C_{14}H_{23}N_5O_4$: C, 51.68; H, 7.13; N, 21.52. Found: C, 51.20; H, 7.08; N, 21.39 %.

15.3.1.2 SYNTHESIS OF 2, 4-DIMETHOXY-6-(PIPERAZIN-1-YL)-1,3,5-TRIAZINE (3D)

3C   3D

Design, Synthesis, and Studies of Novel Piperidine 249

15.3.1.2.1 Procedure

To the solution of tert-butyl 4-(4,6-dimethoxy-1,3,5-triazin-2-yl) piperazine-1-carboxylate (3C) (0.01 mole, 3.25) in DCM was added trifluoro acetic acid (TFA; 0.02 mole, 2.28 g) at 0°C then stirred it at room temperature for 3–4 h. The progress of the reaction was monitored by TLC (10% MeOH/DCM). The solvent in reaction mixture was evaporated to give a sticky mass which was diluted with sat. $NaHCO_3$ and extracted with ethyl acetate. Organic layer was separated, dried over anhydrous sodium sulfate and filtered. Filtrate was concentrated under reduced pressure to get crude product, which upon trituration in n-hexane gives title compound (3D) as off-white solid. The solid compound obtained was dried under vacuum.

15.3.1.2.2 Physicochemical Data

Yield: 62% Mol. Weight: 225.12 Mol. Formula: $C_9H_{15}N_5O_2$

MS(APCI) m/z: 226 [M+1]$^+$.

Anal. Calculated for $C_9H_{15}N_5O_2$: C, 47.99; H,6.71; N, 31.90. Found: C, 47.70; H, 6.40; N, 31.39 %.

15.3.1.3 *GENERAL PROCEDURE FOR THE SYNTHESIS OF 4-(4,6-DIMETHOXY-1,3,5-TRIAZIN-2-YL)-N-SUBSTITUTED PHENYLPIPERAZINE-1-CARBOXAMIDE UREA DERIVATIVES (3A–H)*

250 Green Chemistry and Biodiversity: Principles, Techniques, and Correlations

15.3.1.3.1 Procedure

To the solution of 2, 4-dimethoxy-6-(piperazin-1-yl)-1,3,5-triazine (3D) (0.001 mole, 0.225 g), K_2CO_3 (0.001 mole, 0.138 g) in DMF (for 1g 25 mL) and substituted Isocynate (0.001 mole, 0.133 g) was added at room temperature. The resulting mixture was stirred at room temperature for 2–3 h and the progress of reaction was monitored by TLC (70% EtOAc/Hexane) and (2% MeOH/DCM). On completion of reaction, water was added to give white precipitate which on filtration gives solid. The resulting solid compound (3a–h) was washed with diethyl ether to give title compound which was dried under vacuum.

The analytical, IR and ^{1}H NMR spectral data of 4-(4,6-dimethoxy-1,3,5-triazin-2-yl)-N-substituted phenyl piperazine-1-carboxamide urea derivatives is given in Tables 15.1–15.3.

TABLE 15.1 Analytical and Mass Spectral Data of 4-(4,6-dimethoxy-1,3,5-triazin-2-yl)-N-substituted Phenylpiperazine-1-carboxamide Urea Derivatives 3(a–h).

S. no.	Molecular formula	Molecular weight	Elemental analysis found (calculated)			Mass (m/z)	M.P °C	Yield %
			% C	% H	% N			
3a	$C_{17}H_{22}N_6O_3$	358	56.80 (56.97)	6.19 (6.08)	23.45 (23.39)	359 [M+1]$^+$	127–129	62%
3b	$C_{16}H_{19}N_6O_3Cl$	378	50.63 (50.73)	4.98 (5.06)	22.09 (22.19)	379 [M$^+$]$^+$ 381.81 [M+2]$^+$	140–142	65%
3c	$C_{16}H_{19}N_6O_3F$	362	53.00 (53.03)	5.18 (5.29)	23.09 (23.19)	363 [M+1]$^+$	150–152	66%
3d	$C_{16}H_{19}N_7O_5$	344	49.23 (49.35)	4.85 (4.92)	25.09 (25.18)	390.3 [M+1]$^+$	147–149	60%
3e	$C_{16}H_{20}N_6O_3$	358	55.73 (55.80)	5.70 (5.85)	24.31 (24.40)	345 [M+1]$^+$	123–125	63%
3f	$C_{16}H_{18}N_6O_3F_2$	380	50.41 (50.52)	4.65 (4.77)	22.01 (22.11)	381 [M+1]$^+$	158–160	67%

Design, Synthesis, and Studies of Novel Piperidine   251

TABLE 15.1   *(Continued)*

S. no.	Molecular formula	Molecular weight	Elemental analysis found (calculated)			Mass (m/z)	M.P °C	Yield %
			% C	% H	% N			
3g	$C_{17}H_{22}N_6O_4$	374	54.44 (54.54)	5.77 (5.92)	22.31 (22.39)	375.3 [M+1]$^+$	131–133	63%
3h	$C_{16}H_{20}N_6O_4$	360	53.24 (53.33)	5.47 (5.59)	23.27 (23.32)	361[M+1]$^+$	118–120	62%

TABLE 15.2   IR Spectral Data of 4-(4,6-dimethoxy-1,3,5-triazin-2-yl)-*N*-substituted Phenyl Piperazine-1-carboxamide Urea Derivatives 3(a–h).

S. no.	Substituent (R)	IR band position (wave number cm^{-1})					
		–NH str. in amide	Ar–C–H str. Aromatic	C–H str. in –CH$_3$	>C=O str./ Ar–C=C	C–O–C Ether	–C–N– str. in triazine
3a	4-Methyl	3305	3033	2810	1661(>C=O)/ 1477, 1556	1263	810, 1331
3b	4-chloro	3350	3045	2895	1662/1480, 1560	1255	809, 1361
3c	3-fluoro	3283	3038	2953	1670/1460, 1570	1251	809, 1352
3d	4-Nitro	3257	3142	2951	1662/1458, 1547	1260	811, 1352
3e	phenyl	3275	3086	2850	1645/1469, 1510	1265	812, 1340
3f	2,4-fluoro	3274	3021	2871	1647/1470, 1502	1257	810, 1367
3g	4-methoxy	3280	3040	2876	1647/1440, 1525	1238	813, 1361
3h	4-Hydroxy	3285	3060	2886	1658/1453, 1549	1245/3586	812, 1355

TABLE 15.3 ¹H NMR Data of 4-(4,6-dimethoxy-1,3,5-triazin-2-yl)–*N*-substituted Phenyl Piperazine-1-carboxamide Urea Derivatives (3a–h).

S. no.	R	Assignments, number of proton, multiplicity, and chemical shift in ppm					
		–2OCH₃ in triazine/ Ar-OCH₃	–CH₂–CH₂–/–CH₂– in piperazine	–CH₂–CH₂– in piperazine	Ar–H in Aromatic	–NHAr	Ar–CH₃
3a	4-Methyl	6H, m (3.84)	4H, m (3.64)	4H, m (4.05)	2H, m, (7.59), 2H, m, (7.50)	1H, s (10.61)	3H, s (2.20)
3b	4-Chloro	6H, s (3.87)	4H, m (3.76)	4H, m (3.60)	2H, m, (7.33), 2H, m, (6.92)	1H, s (10.31)	–
3c	3-Fluoro	6H, s (3.76)	2H, m (3.25) 2H, m (4.10)	4H, m (3.72)	2H, m, (7.20), 1H, d, (7.08), 1H, d, (6.93)	1H, s (9.91)	–
3d	4-Nitro	6H, s (3.88)	2H, m (4.46), 2H, m (4.50)	4H, m (3.64)	2H, m (7.50), 2H, m (7.29)	1H, s (10.31)	–
3e	Phenyl	6H, s (3.71)	4H, m (2.84)	4H, m (4.23)	2H, m (7.75), 1H, d (7.59), 2H, m (7.48)	1H, s (10.01)	–
3f	2,4-difluoro	6H, s (3.55)	2H, m (3.19), 2H, m (2.97)	4H, m (3.05)	1H, m (7.14), 1H, d (6.84), 1H, m (6.71)	1H, s (10.12)	–
3g	4-Methoxy	6H, s (3.74) 3H, s (3.45)	2H, m (2.88) 2H, m (2.61)	4H, m (4.06)	1H, m (7.54), 2H, d (7.16), 1H, m (7.11)	1H, s (10.22)	–
3h	4-hydroxy	6H, s (3.85)	2H, m (2.88) 2H, (2.72)	4H, m (4.01)	1H, m (7.47), 2H, d (7.18), 1H, m (6.96)	1H, s (10.25)	Ar-OH 1H, s (4.88)

Design, Synthesis, and Studies of Novel Piperidine 253

15.3.1.4 GENERAL PROCEDURE FOR THE SYNTHESIS OF (4-(4,6-DIMETHOXY-1,3,5-TRIAZIN-2-YL) PIPERAZIN-1-YL) (SUBSTITUTED PHENYL) METHANONE AMIDE DERIVATIVES (3I–P)

15.3.1.4.1 Procedure

To the solution of 2, 4-dimethoxy-6-(piperazin-1-yl)-1,3,5-triazine (3D) (0.001 mole, 0.225 g), K_2CO_3 (0.01 mole, 0.138 g) in DMF (for 1g 30 mL) and 4-methyl benzoyl chloride (0.001 mole, 0.154 g) was added at room temperature. The resulting mixture was stirred at room temperature for 12 h. The progress of reaction was monitored by TLC (70% EtOAc/Hexane) and (2% MeOH/DCM). On completion of reaction, the sufficient water was added and the compound was extracted with ethyl acetate. The organic layer was washed with aq. $NaHCO_3$ solution and water separated out. The product was dried over sodium sulfate and filtered and then concentrated to give solid compound which is recrystallized with diethyl ether to give pure compound (3i–p). It was dried under vacuum.

The analytical data, IR, and ^1H NMR spectral data of (4-(4,6-dimethoxy-1,3,5-triazin-2-yl) piperazin-1-yl) (substituted phenyl) methanone amide derivatives are given in Tables 15.4–15.6.

TABLE 15.4 Analytical and Mass Spectral Data of (4-(4,6-dimethoxy-1,3,5-triazin-2-yl) piperazin-1-yl) (Substituted phenyl) Methanone Amide Derivatives (3i–p).

S. no.	Molecular formula	Molecular weight	Elemental analysis found (calculated)			Mass (m/z)	M.P °C	Yield %
			%C	%H	%N			
3i	$C_{17}H_{21}N_5O_3$	343	59.39 (59.46)	6.10 (6.16)	20.35 (20.40)	344[M+1]$^+$	130–132	66%
3j	$C_{16}H_{18}N_5O_3Cl$	363	52.79 (52.82)	4.87 (4.99)	19.15 (19.25)	364[M+1]$^+$	165–167	66%

Green Chemistry and Biodiversity: Principles, Techniques, and Correlations

TABLE 15.4 *(Continued)*

S. no.	Molecular formula	Molecular weight	Elemental analysis found (calculated)			Mass (m/z)	M.P °C	Yield %
			%C	%H	%N			
3k	$C_{16}H_{18}N_5O_3F$	347	55.25 (55.35)	5.18 (5.22)	20.05 (20.16)	348.9[M+1]$^+$	170–172	66%
3l	$C_{16}H_{18}N_6O_5$	374	51.25 (51.33)	4.78 (4.85)	22.35 (22.45)	375.3[M+1]$^+$	160–162	66%
3m	$C_{16}H_{19}N_5O_3$	329	58.25 (58.35)	5.78 (5.81)	21.15 (21.26)	330[M+1]$^+$	140–142	71%
3n	$C_{16}H_{17}N_5O_3F_2$	365	52.55 (52.60)	4.58 (4.69)	19.05 (19.17)	366[M+1]$^+$	180–182	66%
3o	$C_{17}H_{21}N_5O_4$	359	56.75 (56.82)	5.78 (5.89)	19.39 (19.49)	360.3[M+1]$^+$	150–152	66%
3p	$C_{16}H_{19}N_5O_4$	345	55.55 (55.64)	5.48 (5.55)	20.19 (20.28)	346[M+1]$^+$	157–159	66%

TABLE 15.5 IR Spectral Data of (4-(4,6-dimethoxy-1,3,5-triazin-2-yl) piperazin-1-yl) (Substituted phenyl) Methanone Amide Derivatives (3i–p).

S. no.	Substituent (R)	IR band position (wave number cm^{-1})					
		>C=O str. in	Ar–C–H str. aromatic	C–H str. in –CH$_3$	Ar–C=C- Skelton Str.	C–O–C Ether/– OH str.	–C–N– str. in triazine
3i	4-Methyl	1637	3050	2955	1460, 1547	1249	812, 1356
3j	4-chloro	1684	3015	2964	1484, 1584	1224	813, 1348
3k	3-fluoro	1670	3138	2953	1482, 1570	1251	809, 1352
3l	4-Nitro	1672	3050	2966	1465, 1557	1256	811, 1340
3m	phenyl	1640	3075	2950	1445, 1553	1260	813, 1360
3n	2,4-fluoro	1676	3065	2946	1455, 1584	1252	810, 1360
3o	4-methoxy	1643	3028	2930	1450, 1540	1255	811, 1345
3p	4-Hydroxy	1639	3030	2933	1447, 1559	1261/ 3459	806, 1343

Design, Synthesis, and Studies of Novel Piperidine 255

TABLE 15.6 ^{1}H NMR Data of (4-(4,6-dimethoxy-1,3,5-triazin-2-yl) piperazin-1-yl) (Substituted phenyl) Methanone Amide Derivatives (3i–p).

S. no.	R	Assignments, no. of proton, multiplicity, and chemical shift in ppm				
		$-2OCH_3$ in triazine/Ar	$-CH_2-CH_2-$ in piperazine	$-CH_2-CH_2-$ in piperazine	Ar–H in Aromatic	Ar–CH$_3$
3i	4-Methyl	6H, m (3.84)	4H, m (3.64)	4H, m (4.05)	2H, m, (7.59), 2H, m, (7.50)	3H, s (2.20)
3j	4-Chloro	6H, s (3.87)	4H, m (3.76)	4H, m (3.60)	2H, m, (7.33), 2H, m, (6.92)	–
3k	3-Fluoro	6H, s (3.76)	2H, m (3.25) 2H, m (4.10)	4H, m (3.72)	2H, m, (7.20), 1H, d, (7.08), 1H, d, (6.93)	
3l	4-Nitro	6H, s (3.88)	2H, m (4.46) 2H, m (4.50)	4H, m (3.64)	2H, m (7.50), 2H, m (7.29)	–
3m	Phenyl	6H, s (3.71)	4H, m (2.84)	4H, m (4.23)	2H, m (7.75), 1H, d (7.59), 2H, m (7.48)	–
3n	2,4-difluoro	6H, s (3.55)	2H, m (3.19), 2H, m (2.97)	4H, m (3.05)	1H, m (7.14), 1H, d (6.84), 1H, m (6.71)	
3o	4-Methoxy	6H, s (3.74), 3H, s (3.45)	2H, m (2.88), 2H, m (2.61)	4H, m (4.06)	1H, m (7.54), 2H, d (7.16), 1H, m (7.11)	–
3p	4-hydroxy	6H, s (3.85)	2H, m (2.88), 2H, m (2.72)	4H, m (4.01)	1H, m (7.47), 2H, d (7.18), 1H, m (6.96)	Ar–OH 1H, s (4.88)

15.4 RESULTS AND DISCUSSION

The IR spectra of the compounds (3a–h) and (3i–p) showed the characteristic stretching bands at 3300–3350 cm^{-1} (–NH), 1650–1690 cm^{-1} (>C=O), 1200–1263 cm^{-1} (–C–O–C– in ether) and 1300–1331 cm^{-1} (–C–N– in triazine). The ^{1}H NMR spectra of the compounds (3a–h) and (3i–p) showed the characteristic peaks at δ 10.01–10.61 (S) due to the presence of –NH proton and δ 3.01–4.23 (m) due to the presence of piperazine rings. The mass spectra of synthesized products are in agreement with their molecular weights.

15.5 BIOLOGICAL EVALUATION

15.5.1 INTRODUCTION

1,3,5-triazine, a six-membered heterocyclic rings, is one of the oldest classes of organic compound. It is still used in many chemotherapeutics agents due to their interesting pharmacological properties, including anticancer, herbicidal, insecticidal, anti-HIV, anti-malarial, antibacterial, antimycobacterial, and antimicrobial activities.[18]

The cytokines are intercellular messengers accountable for mass defense mechanisms as inflammatory, immune, and hematogenic responses. While numerous of them are transient, produced by a variety of cells acting as insistent response mediators in cases of invasive interventions. The Interference of this biological resistance mechanism and constant extreme cytokine production contributes to pathogenesis of inflammatory diseases. The tumor necrosis factor alpha (TNF-α) and interleukin-6 (IL-6) are the two key multifunctional pro-inflammatory cytokines, occupied in the pathogenesis of inflammatory, cancer diseases, neurodegenerative, autoimmune, and cardiovascular during a sequence of cytokine signaling pathways by Papadakis et al.[19] The most important type pro-inflammatory cytokine TNF-α is a huge number of biological activities associated to pathology of autoimmune diseases such as rheumatoid arthritis (RA),[20] Crohn's disease,[21] systemic lupus erythematosus,[22] and multiple sclerosis,[23] and septic shock.[24] Alternatively, Dominic et al. revealed that, integral role of cytokine interleukin-6 in the pathogenesis of anemia of chronic disease suggests that it could be an important therapeutic target. Currently, available treatments target interleukin-I and tumor necrosis factor-α and its receptors and have been only partially successful.[25]

TNF-α and IL-6 are the most biological and pharmaceutically essential molecular focal point for the action of the above mentioned diseases. The existing biopharmaceuticals (TNF soluble receptor (Enbrel TM) and TNF antibody (Remicade TM) are luxurious, complicated to administer orally, and have major side effects on lengthened medical

Design, Synthesis, and Studies of Novel Piperidine 257

use. Consequently, there is a critical therapeutic requirement to find out little molecule agents to deal with more production of TNF-α. Triazine is found in many potent biologically active molecules with promising biological potential like anti-inflammatory, anti-mycobacterial, anti-viral, anti-cancer, and so on, which makes it an attractive scaffold for the design and development of new drugs.[26] It is our interest to investigate that hybrid compound incorporating a piperazine core into triazine nucleus with urea/amide derivatives in single molecular frame could lead to the novel potent anti-inflammatory and antimicrobial agents. Hence, we report the synthesis, anti-inflammatory and antimicrobial activity evaluation study of novel 4-(4,6-dimethoxy-1,3,5-triazin-2-yl)-N-phenyl piperazine-1-carboxamide urea derivatives (3a–h) and (4-(4,6-dimethoxy-1,3,5-triazin-2-yl) piperazin-1-yl) (phenyl)methanone amide derivatives (3i–p).

15.5.2 ANTI-INFLAMMATORY ASSAY

Pro-inflammatory cytokine production by lip polysaccharide (LPS) in THP-1 cells was measured according to the method described by Hwang et al.[27] During the assay, THP-1 cells were cultured in RPMI 1640 culture medium (Gibco BRL, Pasley, UK) containing 100 U/mL Penicillin and 100 mg/mL Streptomycin containing 10% fetal bovine serum (FBS, JRH). Cells were differentiated with phorbol myristate acetate (PMA, Sigma). Following cell plating, the test compounds 4-(4,6-dimethoxy-1,3,5-triazin-2-yl)-N-phenylpiperazine-1-carbox-amide urea derivatives (3a–h) and (4-(4,6-dimethoxy-1,3,5-triazin-2-yl)piperazin-1-yl)(phenyl)methanone amide derivatives (3i–p) in 0.5% DMSO were added to each well separately and the plate was incubated for 30 min at 37°C. Finally, LPS (*E. coli* 0127:B8, Sigma Chemical Co., St. Louis, MO) was added, at a final concentration of 1 µg/mL in each well. Plates were further incubated at 37°C for 24 h in 5% CO_2. After incubation, supernatants were harvested, and assayed for TNF-α and IL-6 by ELISA as described by the manufacturer (BD Biosciences). The % inhibitions are measured at 10 µM concentration. The Dexamethasone is used as standard drug. The results are tabulated in Tables 15.7 and 15.10.

15.5.3 ANTIBACTERIAL ASSAY

Newly synthesized compounds were screened for their antibacterial activity against selected Gram-positive organism's, namely, *Staphylococcus aureus* (MTCC 96), *Bacillus subtilis* (MTCC 441), and Gram-negative organism's, namely, *Escherichia coli* (MTCC 443), *Salmonella typhimurium* (MTCC 98) bacterial strains by agar well diffusion method with little modification.[28] The different concentrations (10–200 µg/mL) of test compounds 4-(4,6-dimethoxy-1,3,5-triazin-2-yl)-*N*-phenylpiperazine-1-carbox-amide urea derivatives (3a–h) and (4-(4,6-dimethoxy-1,3,5-triazin-2-yl)piperazin-1-yl)(phenyl)methanone amide derivatives (3i–p) were prepared in DMSO. The bacterial suspension was spread over nutrient agar plates and the well with 6 mm diameter was punched with sterile cork borer. The sample (3a–h) and (3i–p) (50 µL) was added to the well and the plates were incubated at 37°C for 24 h. Respective solvent control (DMSO) was kept and ciprofloxacin was used as standard antibacterial agent. The lowest concentration of compound which completely inhibits the bacterial growth was taken as minimum inhibitory concentration (MIC) and the MICs were noted for antibacterial assay. The results are tabulated in Tables 15.8 and 15.11.

15.5.4 ANTIFUNGAL ASSAY

Newly synthesized compounds 4-(4,6-dimethoxy-1,3,5-triazin-2-yl)-*N*-phenylpiperazine-1-carbox-amide urea derivatives (3a–h) and (4-(4,6-dimethoxy-1,3,5-triazin-2-yl)piperazin-1-yl)(phenyl)methanone amide derivatives (3i–p) were screened for their antifungal activity against *Candida albicans* (MTCC 227), *Aspergillus niger* (MTCC 281), *Fusarium solani* (MTCC 350), and *Aspergillus flavus* (MTCC 277) by agar well diffusion method with little modification Sridhar et al.[28] Normal saline was used to make a suspension of spores of fungal strain. The fungal suspension was spread over potato dextrose agar plates and the wells of 6 mm diameter were punched with sterile cork borer. The different concentrations (10–200 µg/mL) of test compounds were prepared in DMSO. The sample 3a–h and 3i–p (50 µL) was added to the well and the plates were incubated at 37°C for 2 to 3 days. Respective solvent control (DMSO) was kept and miconazole was used as standard antifungal agent. The lowest concentration of compound which completely inhibits the fungal growth

Design, Synthesis, and Studies of Novel Piperidine 259

was taken as MIC and the MICs were noted for antifungal assay. The results are tabulated in Tables 15.9 and 15.12.

15.5.5 SAR OF SYNTHESIZED COMPOUNDS

The series of the novel 4-(4,6-dimethoxy-1,3,5-triazin-2-yl)-N-substituted phenyl piperazine-1-carboxamide urea derivatives (3a–h) and (4-(4,6-dimethoxy-1,3,5-triazin-2-yl)piperazin-1-yl)(substituted phenyl) methanone amide derivatives (3i–p), were evaluated for their ability to inhibit pro-inflammatory cytokines (TNF-α and IL-6) activity by TNF-α and IL-6 inhibition assay and anti-microbial activity against various gram-positive, gram-negative bacteria and fungal strains by using an agar well diffusion method. As presented in Tables 15.7–15.12, the tested compounds demonstrated a wide range of potencies which clearly showed the contributions of the urea/amide triazine structure in terms of structure–activity relationships (SAR).

Thus, from the TNF-α and IL6 inhibitory activity data (Tables 15.7 and 15.10), it is observed that the compounds 3c, 3f, 6k, and 6n found to be highly active as TNF-α and IL-6 inhibitor (up to 65–73% TNF-α and 73–85% IL-6 inhibitory activity) with compounds 3f (73% and 85%) and 3n (69% and 78%) exhibiting highest inhibition against TNF-α and IL-6, respectively at 10 μM. It is to be noted that almost these entire compounds either found to be equally potent or more potent than that of the standard dexamethasone at 1 μM. The compounds 3b, 3g, 3o, and 3p exhibited moderate activity (45–60% inhibition) while the other compounds 3a, 3d, 3e and 3i, 3l, 3m exhibited low to very low or no activity all at same level of concentration. As shown in Tables 15.7 and 15.10, we firstly introduced urea and amide functionality on the triazine ring and a comparison of different substitutions at the position R_1 and R_2 for their activity has been studied. The SAR with respect to the anti-inflammatory activity for the compounds (3a–h) and (3i–p) have shown an interesting trend as to the effect of substituent present on terminal ring of urea and amide moiety as well as the electronic nature of piperazine moiety in it has shown profound effect on the activity.

It is found from our results Tables 15.8 and 15.11 that the nature and position of substituent on terminal benzene ring of urea and amide moiety has significant effect on the biological activity. When fluorine (F) atom is present on 2, 3, and 4 positions on the terminal benzene ring of urea and amide moiety, it is found that these positions are the favorable sites for the higher potency. Evidently, the compound 3f and 3n with –F at 2 and 4 positions on the terminal benzene rings of urea and amide series exhibiting highest TNF-α and IL-6 inhibitory activity. Also the presence of –F atom at position 3 of terminal benzene ring have shown the good activity (3c, 3k). While the presence of Cl at 4 and 2- position of terminal benzene ring 3b, and 3j, respectively and they exhibit moderate (47–59%) inhibitory activity. Interestingly, the large bulky group such as Methyl and –NO$_2$ at 4-position of terminal benzene ring generated compound 3a, 3d, 3i, and 3l from both series has no effect on TNF-α or IL-6 inhibitory activity. The –OH group in amide series with a moderate activity, whereas for urea series it shows less activity.

It reveals from SAR studies that, the presence of –F tolerates the procytokine activity because it is found that fluorine imparts the special characteristics that enhance therapeutic efficacy and improved pharmacological properties in bioactive molecules. Thus, the compounds 3f and 3n were found to be most potential inflammatory agents amongst the series of compounds studied and could prove to be promising candidate for drug discovery. Though there is no actual evidence in hand in support of the actual role of urea or amide moiety on the activity at this time, so it is speculated that the H-bond formation ability of the urea or amide along with the electronic effect of para substituent might be responsible for their high anti-inflammatory activity. Explicitly, it is electronic effect of the substituent on piperazine ring which mainly influences the anti-inflammatory activity.[29] As a good response against anti-inflammatory activity shown by above mentioned urea and amide derivatives of (piperazin-1-yl)-1,3,5-triazine. It is found wise to go for the antimicrobial activity of these compounds to further assist for SAR study.

From antimicrobial activity data shown in Tables 15.8, 15.9, 15.11, and 15.12, it is revealed that some analogues of this series have more potency than the standard drug Ciprofloxacin and Miconazole while some of them have comparable potency. Interestingly none of the compound with high anti-inflammatory activity found to be potent antibacterial or antifungal agents. Thus, the compounds 3a, 3d, 3e as well as 3i, 3l, and 3m bearing aryl 4-Me, 4-NO$_2$ and phenyl group respectively of both the series have higher potency against the tested antimicrobial strain. It is cleared from our results that the 4th position of substituent on terminal benzene ring of urea and amide moiety is the favorable site for high antimicrobial

Design, Synthesis, and Studies of Novel Piperidine   261

activity. The high potency of these compounds may be attributed to the presence of H-bond acceptor type group's placement at 4-positions. While presence of Cl, –OH on terminal benzene ring of urea and amide series (3b, 3h, 3j, and 3p) shows moderately potent antimicrobial activity with respect to standard drug. Attractively the presence of –NO$_2$ at 4th position of terminal benzene ring in urea series (3d) shows potent antibacterial and antifungal activity while in amide series (3l) shows moderate antibacterial and antifungal activity. Any activity has not been observed in case of remaining compounds up to concentration of 200 µg/mL against same bacterial and fungal strains. Higher inhibition effects observed dependent on –Cl, –F, –OCH$_3$ functionality to the N-atom of the piperazine bases condensed to nucleus.[30]

It is cleared from results, that is, Tables 15.8 and 15.9 for series of urea versus Tables 15.11 and 15.12 for series of amide that, the SAR of antibacterial activity partially correlates with their SAR of antifungal activity as there is some divergence is observed. The same aryl 4th position as observed already is favorable site of high activity for both series. The compounds 3d and 3m have been found to be 2–2.5-fold more potent than the standard drug Micanazole as similar to the antibacterial activity trend. Only dissimilarity is observed in case of compound 3l. While the compounds (3k, n, and o) have no major effect on the antifungal activity also. Explicitly the compound 3a, 3d, and 3l bearing single substituent along with urea and amide functionality reflect 2.5-fold antimicrobial activity against the pathogenic bacteria and fungal strain.

TABLE 15.7 Anti-inflammatory Activity Data of 4-(4,6-dimethoxy-1,3,5-triazin-2-yl)-N-substitutedphenylpiperazine-1-carboxamide Urea Derivatives (3a–h).

Comp. (3a–h)	(R$_1$)	% Inhibition at 10 µM	
		TNF-α	IL-6
3a	4-Methyl	15	16
3b	4-Chloro	55	59
3c	3-fluro	67	73
3d	4-NO$_2$	25	28
3e	Phenyl	10	13
3f	2,4-difluoro	73	85
3g	4-Methoxy	53	61
3h	4-Hydroxy	30	34
Ref.	Dexamethasone (1µM)	75	84

TABLE 15.8 Antibacterial activity of 4-(4,6-dimethoxy-1,3,5-triazin-2-yl)-*N*-substitutedphenylpiperazine-1-carboxamide Urea Derivatives (3a–h).

Comp. (3a–h)	Gram-positive (MIC[a] values µg/mL)		Gram-negative (MIC values µg/mL)	
	S. aureus	*B. subtilis*	*E. coli*	*S. typhimurium*
3a	10	15	10	10
3b	40	30	45	50
3c	80	65	80	80
3d	15	10	10	10
3e	20	15	15	20
3f	80	90	00	85
3g	70	55	65	75
3h	65	60	55	65
Ciprofloxacin (Ref.)	25	25	15	25

[a]Values are the average of three reading.

TABLE 15.9 Antifungal Activity of 4-(4,6-dimethoxy-1,3,5-triazin-2-yl)-*N*-substituted phenyl piperazine-1-carboxamide Urea Derivatives (3a–h).

[(MIC[a] values µg/mL)]

Comp. (3a–h)	*Candida albicans*	*Aspergillus niger*	*Fusarium – solani*	*Aspergillus flavus*
3a	15	25	10	10
3b	30	30	35	40
3c	75	65	95	80
3d	10	20	10	10
3e	45	30	40	35
3f	90	90	00	00
3g	95	75	85	90
3h	70	65	60	65
Miconazole (Ref.)	20	25	15	15

[a]Values are the average of three readings.

Design, Synthesis, and Studies of Novel Piperidine 263

TABLE 15.10 Anti-inflammatory Activity Data of (4-(4,6-dimethoxy-1,3,5-triazin-2-yl) piperazin-1-yl) (substituted phenyl)Methanone Amide Derivatives (3i–p).

Comp. (3i–p)	(R₂)	% Inhibition at 10 μM	
		TNF-α	IL-6
3i	4-Methyl	21	25
3j	2-Chloro	47	59
3k	3-fluro	65	74
3l	4-NO$_2$	31	27
3m	Phenyl	0	10
3n	2,4-difluoro	69	78
3o	4-Methoxy	49	54
3p	4-hydroxy	43	47
Ref.	Dexamethasone (1 μM)	75	84

TABLE 15.11 Antibacterial Activity of (4-(4,6-dimethoxy-1,3,5-triazin-2-yl)piperazin-1-yl) (substituted phenyl) Methanone Amide Derivatives (3i–p).

[(MIC[a] values μg/mL)].

Comp. (3i–p)	Gram-positive MIC values μg/mL)		Gram-negative (MIC[a] values μg/mL)	
	S. aureus	B. subtilis	E. coli	S. typhimurium
3i	20	15	15	20
3j	60	65	70	65
3k	80	65	80	90
3l	30	40	45	35
3m	25	15	20	10
3n	75	95	00	85
3o	70	65	65	75
3p	55	50	60	75
Ciprofloxacin (Ref.)	20	25	20	15

[a]Values are the average of three readings.

TABLE 15.12 Antifungal Activity of (4-(4,6-dimethoxy-1,3,5-triazin-2-yl)piperazin-1-yl)(substituted phenyl) Methanone Amide Derivatives (3i–p).

[(MIC^a values µg/mL)].

Comp. (3i–p)	Candida albicans	Aspergillus niger	Fusarium - solani	Aspergillus flavus
3i	25	15	15	20
3j	65	75	80	65
3k	80	65	0	80
3l	35	40	45	45
3m	25	15	20	10
3n	90	90	0	0
3o	95	80	85	90
3p	65	70	60	80
Miconazole (Ref.)	20	25	20	15

^{a}Values are the average of three readings.

15.6 CONCLUSION

We have successfully synthesized the (piperazin-1-yl)-1,3,5-triazine-based novel 4-(4,6-dimethoxy-1,3,5-triazin-2-yl)-N-substituted phenylpiperazine-1-carboxamide urea derivatives (3a–h) and (4-(4,6-dimethoxy-1,3,5-triazin-2-yl)piperazin-1-yl)(substituted phenyl) methanone amide derivatives (3i–p). In the overall reaction protocol present an atom economy scheme as HCl and CO_2, and so on is the only byproduct. The process offers simple reaction profile, good yield 62–67% for carboxamide urea derivatives followed by 66–71% for methanone amide derivatives and easy work up procedure. Except first step, all reactions are carried out at room temperature which presents itself an additional environmentally benign synthetic advantage. The 16 derivatives (3a–h and 3i–p) were synthesized and characterized for their structure elucidation by using physical, analytical, and Spectral data (NMR, Mass).

All compounds were evaluated for their anti-inflammatory (against TNF-α and IL-6) as well as antimicrobial activities against different gram positive and gram negative bacterial and fungal strains. Among all the compounds screened (3a–h and 3i–p), the compounds 3c, 3f, 3k, and 3n showed promising TNF-α and IL-6 inhibitory activity with compounds 3f and 3n exhibiting highest activity while the compounds 3a, 3d, 3e, 3i,

Design, Synthesis, and Studies of Novel Piperidine 265

3l, and 3m found to be the potent antimicrobial agent, showing even 2- to 2.5-fold more activity than that of standard Ciprofloxacin and Miconazole at the same MIC value of 10 µg/mL. Thus the presence of chlorine and fluorine tolerates the procytokine activity as well as the H-bond donor ability of the urea and amide might be responsible for their high anti-inflammatory activity. While the presence of Methyl, $-NO_2$, and phenyl group on 4th position of terminal benzene ring of urea and amide functionality found to be effective potent antimicrobial agents.

KEYWORDS

- **amide**
- **antimicrobial activity**
- **anti-inflammatory activity**
- **aryl urea**
- **piperazine**
- **triazine**

REFERENCES

1. (a) Simanek, E. E.; Abdou, H.; Lalwani, S.; Lim, J.; Mintzer, M.; Venditro, V. J.; Vittur, B. *Proc. R. Soc. A* **2010,** *466,* 1445–1451. (b) Inca, S. Z.; Selma, S.; Semra, C.; Kevser, E. *Bioorg. Med. Chem.* **2006,** *14,* 8582–8586.
2. Zacharie, B.; Abbott, S. D. *J. Med. Chem.* **2010,** *53,* 1138–1142.
3. Singh, K. K.; Joshi, S. C.; Mathela, C. S. *Ind. J. Chem.* **2011,** *50 B,* 196–200.
4. Anandakumar, C. S.; Vinaya, K.; Sharathchandra, J. N.; Thimmegowda, N. R.; Benkaprasad, S. B.; Sadashiva, C. T.; Rangappa, K. S. *J. Enz. Inhib. Med. Chem.* **2008,** *23,* 462–465.
5. Patel, R. V.; Kumari, P.; Chikhalia, K. H. *Arch. Appl. Sci. Res.* **2010,** *2,* 232.
6. Sarmah, K. N.; Sarmah, N. K.; Patel, T.V. *J. Chem. Pharm. Res.* **2014,** *6* (9), 127–132.
7. Kerns, R. J.; Rybak, M. J.; Kaatz, G. W.; Vaka, F.; Cha, R.; Grucz, R. G.; Diwadkar, V. U. *Bioorg. Med. Chem. Lett.* **2003,** *13,* 2109–2113.
8. Ryckebusch, A.; Poulain, R.; Maes, L.; Debreu-Fontaine, M. A.; Mouray, E.; Grellier, P.; Sergheraert, C. *J. Med. Chem.* **2003,** *46,* 542–545.
9. Becker, J. *J. Heterocyc. Chem.* **2008,** *45,* 1005–1009.
10. Kimura, M.; Masuda, T.; Yamadaa, K.; Mitania, M.; Kubota, N.; Kawakatsu, N.; Kishii, K.; Inazu, M.; Kiuchi, Y.; Oguchi, K.; Namiki, T. *Bioorg. Med. Chem.* **2003,** *11,* 3953–3955.

11. Chikhalia, K. H.; Lakum, H. P.; Desai, D.V. *Heterocyl. Commun.* **2013**, *19* (5), 351–355.
12. Hackbarth, C. J.; Chen, D. Z.; Lewis, J. G. *Chemother* **2002**, *46*, 2752–2756.
13. Pomarnacko. E.; Bednarski, P.; Grunert, R.; Reszka, P. *Acta Poloniae Pharm. Drugs Res.* **2004**, *61* (6), 461–466.
14. Todd, A. B.; Jennifer, A. M.; Michael, P. C.; Mark, S. *Bioorg. Med. Chem. Lett.* **2006**, *16*, 3510–3515.
15. Jennifer, A. M. *Bioorg. Med. Chem Lett.* **2006**, *16*, 3514–3519.
16. Hiroshi, E.; Ayako, S.; Hiroshi, S.; Noriyoshi, Y.; Hir oyuki, I.; Chikako, S.; Fumio, Masahiro, T.; Yoshimas, O. S.; Masato, H.; Masakazu, B. *Bioorg. Med. Chem. Lett.* **2010**, *20*, 4479–4485.
17. Desai, S. D.; Mehta, A. G. *Res. J. Chem. Sci.* **2014**, *4* (5), 14–18.
18. Shah, D. R.; Modh, R. P.; Chikhalia, K. H. *Future Med. Chem.* **2014**, *6* (4), 463–466.
19. Papadakis, K. A.; Targan, S. R. *Inflamm. Bowel. Dis.* **2000**, *6*, 303–308.
20. Macnaul, K. L.; Hutchinson, N. I.; Parsons, J. N.; Bayne, E. K.; Tocci, M. J. *J. Immunol.* **1990**, *145*, 4154–4160.
21. Dullemen, H. M. V.; Deventer, S. J. H. V.; Hommes, D. W.; Bijl, H. A.; Jansen, J.; Tytgat, G. N. J.; Woody, *J. Gastroenterol.* **1995**, *109*, 129–133.
22. Maury, C. P. J.; Teppo, A. M. *Arthr. Rheum.* **1989**, *32*, 146–150.
23. (a) Sharief, M. K.; Hentges, R. N. *Engl. J. Med.* **1991**, *325*, 467–471. (b) Beck, J.; Rondot, P.; Catinot, L.; Falcoff, E.; Kirchner, H.; Wietzerbin, J. *Acta Neurol. Scand.* **1988**, *78*, 318–322.
24. Lechner, A. J.; Tredway, T. L.; Brink, D. S.; Klein, C. A.; Matuschak, G. M. *Am. J. Physiol.* **1992**, *263*, 526–530.
25. Dominic, S. C.; Raj, M. D. Semin. *Arthritis Rheum.* **2009**, *38*, 382–386.
26. Singla, P.; Luxami, V.; Paul, K. *Eur. J. Med. Chem.* **2015**, *102*, 39–45.
27. Hwang, C.; Gatanaga, M.; Granger, G. A.; Gatanaga, T. *J. Immunol.* **1993**, *151*, 5631–5636.
28. Sridhar, R.; Perumal, P. T.; Etti, S.; Shanmugam, G.; Ponnuswamy, M. N.; Prabavathyc, V. R.; Mathivanan, N. *Bioorg. Med. Chem. Lett.* **2004**, *14*, 6035–6040.
29. Hatnapure, G. D.; Keche, A. P. et al. *Bioorg. Med. Chem. Lett.* **2012**, *22*, 6385–6390.
30. Patel, R. V.; Kumari, P.; Rajani, D. P.; Chikhalia, K. H. *Eur. J. Med. Chem.* **2011**, *46*, 4353–4558.

CHAPTER 16

Metaphors That Made History: Reflections on Philosophy/Science/DNA

FRANCISCO TORRENS[1*] and GLORIA CASTELLANO[2]

[1]*Institut Universitari de Ciència Molecular, Universitat de València, Edifici d'Instituts de Paterna, P. O. Box 22085, E46071 València, Spain*

[2]*Departamento de Ciencias Experimentales y Matemáticas, Facultad de Veterinaria y Ciencias Experimentales, Universidad Católica de Valencia San Vicente Mártir, Guillem de Castro-94, E46001 València, Spain*

Corresponding author. E-mail: torrens@uv.es

ABSTRACT

Research in two-dimensional nanomaterials augurs electronics-world revolution. Keys for powerful, miniaturized devices are conceived in chemistry/physics laboratories. How does molecular science contribute to improving two-dimensional-nanomaterials properties? How to handle substance? People familiar with matter manipulation create much civilization stuff. Alchemists, chemists, and molecular scientists seek to expand matter understanding. Frontier science provides hope for problems but inequality persists. Science must contribute to life being better than the previous generation for everybody. To associate word chemical with a scientific laboratory is common. Now, chemical stands juxtaposed to natural. While many manmade chemicals are toxic, the same occurs with natural products. The artificial/natural dilemma comes from alchemy: A cause is lack of spreading. Need for communicating science is a secondary effect of scientific creativity. The principle of caution should be applied.

Subjects come unstrung. What to do if science policy comes into conflict with boss/employees? Bioethics subject is to look after life. One never had as many possibilities of taking care as now. However, life was never as much threatened as today. One can look after life but destroy it with greater easiness. Bioethics becomes ethics of responsible management of human life. Notwithstanding, laic ethics/religions arise, where theological positions must abandon dogmatism and moral monopoly. Scientific questioning creates hypotheses, trusting discoveries. Fiduciary listening dares to wait that revelations become public. Both are capable of searching, questioning, and changing one's mind faced with data.

16.1 INTRODUCTION

Setting the scene: Reflections on metaphors that have made history in philosophy, science, and analogies examples in teaching DNA, building a house, and bread-baking. Working with nanomaterials (NMs), the principle of caution should be applied. Research in two-dimensional (2D) NMs augurs an electronics-world revolution. Keys for powerful, tiny-devices development are conceived in chemistry and physics laboratories throughout the world, end to which a molecular-interactions study is a tool. How does molecular science contribute to improving 2D-NMs properties, facing the development of future electronic devices? Chalcogen nanocomposites (NCs) are analyte preconcentration, imaging, and detection tools from environmental to proteomic studies. People familiar with the matter and how to manipulate it created much civilization stuff. They may not call themselves chemists but their experimentation shaped the modern world. For the sheer joy of knowing, alchemists, chemists, and molecular scientists sought to expand people's matter understanding from the atomic to the galactic. Frontier science provides hope for problems. However, inequality persists. Science must contribute toward life being better than the previous generation for everybody. To associate the word chemical with the scientific laboratory is common, as something arising out of beakers and flasks at the hands of chemists wearing laboratory coats. In 21st century, chemical stands juxtaposed to natural and, increasingly in public opinion, the two are considered mutually exclusive occurrences. While many manmade chemicals are toxic, the same occurs with plant products. The artificial/natural dilemma comes from the times of alchemy: The cause is the lack of spreading.

General bioethical aspects were tackled from historical notes to technique foundations, and concepts anthropological bioethics and law. Subjects came unstrung. American Chemical Society (ACS) informed Academic Professional Guidelines. However, they do not explain what to do if science policy comes into conflict with boss or employees. Medicine first challenge is to solve bioethical problems. Bioethics subject is to look after life. People never had as many possibilities of taking care of this as nowadays, on aspects related to factors, such as hygiene, health, gynecology, pediatrics, geriatrics, transplants, palliatives, genetically modified, etc. Notwithstanding, life was never as much threatened as these days, including wars, violence, unfair sharing goods, technology abuses, environmental destruction, etc. At the present time, people can look after life. However, they can destroy it with greater easiness. Bioethics becomes ethics of responsible management of human life, in the framework of fast biomedical advances, and therapeutic limits or therapy/research frontiers must be rethought. Nevertheless, laic ethics/religious perspectives meeting arises, where theological positions should abandon dogmatism, fundamentalism, and moral monopoly; in theological ethics, one should make possible discrepancy without dissidence versus magisterial authorities, and tradition criticism with creative fidelity. Scientific questioning creates hypotheses, trusting discoveries. Fiduciary listening dares to wait that revelations become public. Both are capable of searching, questioning, and changing one's mind faced with data, without allowing himself to be rubbed off by relativist-fear syndrome. However, ideologies, with science or faith mask, neither doubt nor investigate: They stick to deceptive certainties, longing for securities, and creating pseudoscience/pseudoreligion, which are usually dogmatic-fanaticisms uniform.

In earlier publications, it was reported the periodic table of the elements (PTE),[1-3] quantum simulators,[4-12] science, ethics of developing sustainability via nanosystems, devices,[13] green nanotechnology, as an approach toward environment safety,[14] molecular devices, machines as hybrid organic–inorganic structures,[15] PTE, quantum biting its tail, sustainable chemistry,[16] quantum molecular spintronics, nanoscience, and graphenes.[17] It was informed cancer, its hypotheses,[18] precision personalized medicine from theory to practice, cancer,[19] how human immunodeficiency virus/acquired immunodeficiency syndrome (HIV/AIDS) destroy immune defences, hypothesis,[20] 2014 emergence, spread, uncontrolled Ebola outbreak,[21,22] and Ebola virus disease, questions, ideas, hypotheses, and models.[23] In the present chapter, it is discussed some reflections on some metaphors that have made history in philosophy, science, and analogies examples in teaching DNA: Building a house, and bread-baking.

16.2 SOME METAPHORS THAT HAVE MADE HISTORY IN PHILOSOPHY

Strauss published an introduction to philosophy.[24] Table 16.1 gives some metaphors that have made history in philosophy.[25]

TABLE 16.1   Some Metaphors that Have Made History in Philosophy.

Metaphor
Humans are born to die.
We all die in the end.
(Cervantes). As much true, Sancho, as someday we will be dead.
(Keynes). In the long run, we are all dead.
(Louis L'Amour). The more one learns the more he understands his ignorance.
(Albert Einstein). The more I learn, the more I realize how much I don't know.

16.3 SOME METAPHORS THAT HAVE MADE HISTORY IN SCIENCE

A new tool, clustered, regularly interspaced, short palindromic repeat (CRISPR), allows scientists to modify the genome like editors that rewrite the book of life.[26] This is one of the literary figures about the advances of genetics. Since the Human Genome Project (HGP, 1990), metaphors followed their vital cycle: they are born, settle into, die, and are replaced. Although in the 2000s, DNA was a map to decipher, now it is a book. These images mark the evolution of science. Keller noticed that the ways in which people talk about science objects are not simply determined by empirical evidence, but they actively influence the type of proof that people search and how research is driven.[27] Table 16.2 lists some metaphors that have made history in science. Lakoff and Johnson analyzed how the metaphors that people use to model their perception, thought, and

Metaphors That Made History 271

actions.[28] Sontag noticed that the utilization of war language, in cancer description, can present an unbearable weight in which suffer from it.[29] Lizcano noticed that the problem appears when, because of their repeated application, these metaphors turn invisible, natural.[30] The PTE is related to the literature.[31]

TABLE 16.2 Some Metaphors that Have Made History in Science.

Concept	Metaphor
Electricity	Current
Atom	Solar system in miniature
Nature	Book written in mathematical language
Brain	Human–computer
Mind	Brain software
Amino acids	Bricks of life
Genome	Code, map, manual, identity card, genetic passport
Internet	Information highway
Mitochondria	Cell power plants
Immune system	Defense army
Immune system	Battlefield
Immune system	Arms race
Radiologists that avoid destroying healthy tissue	Soldiers conquering a territory without running into the mines there hidden (Gick and Holyoak, 1980)
Virus	Invading enemy
Antibiotic	Weapon against bacteria
Enzyme and substrate	Lock and key
Drug	Magic bullet
Protein, nucleic acid, etc.	Biological target
Body	Cell society, machine
Heart	Blood pump
Nervous system	Electric net
Earth	Living organism
Elementary particles	Bricks of matter
Space–time	Weave of cosmos
Higgs' boson	Particle of God
CRISPR	Genetic cut–paste, molecular scissors, molecular scalpel

272 Green Chemistry and Biodiversity: Principles, Techniques, and Correlations

16.4 ANALOGIES EXAMPLES IN TEACHING DNA: BUILDING A HOUSE/BREAD-BAKING

The world of literary figures and metaphors is important in science education. Table 16.3 collects some analogies examples in teaching DNA concepts: Building a house and bread-baking. Research in a comic style has opportunities in science education.[32]

TABLE 16.3 Analogies Examples in Teaching DNA Concepts: Building a House and Bread-baking.

Like (where the analogies match the concepts)		
Transcription and translation concepts	Building a house analogy	Bread-baking analogy
DNA-the set of instructions for the cell	Master plan of a house (building plan)	Cookbook (recipe) containing a set of instructions for bread making
Gene-is the single DNA instruction	Is a single instruction/step in the building plan	Is a single instruction/step in the cookbook/recipe
Ribosome	Foreman/contractor	Cook/chef
Cytoplasm	Construction site	Kitchen
Nucleus	Foreman's office at the site	Pantry (store room for keeping ingredients)
Amino acids (AAs)	Blocks	Ingredients
Transcribing DNA to make messenger RNA (mRNA) (i.e., transcription)	Making a copy of the master plan for use at the construction site	Making a copy of the recipe used at the kitchen
mRNA	Photocopy of the master plan	Photocopy of the original recipe plan
The ribosome reads the mRNA one codon at a time	The foreman reads the photocopied plan one step at a time	The chef reads the photocopied recipe one step at a time
Every codon corresponds to a specific AA	Every step in the plan refers to a specific design/style	Every step in the plan refers to a specific ingredient

Metaphors That Made History   273

TABLE 16.3 *(Continued)*

AAs are sent to the ribosome by transfer RNA (tRNA)	Blocks and other materials are sent to the supervisor by truck	Ingredients are sent to the chef by kitchen hands or car
Ribosome joins AAs to form proteins/ polypeptides (i.e., translation)	The foreman follows the plan to lay/arrange the blocks to build a house	The chef mixes ingredients food or bread
Energy to bond AAs to form proteins	Mortar to bond blocks together	–
Protein	Finished building	Finished bread/food
The same kinds of AAs can be arranged differently to build lots of different proteins	The same kinds of blocks can be arranged differently to build lots of different houses	The same ingredients can be mixed/combined differently to make different kinds of bread or food
Mutation	Mistakes made by block foreman	Mistakes made by the chef

Limitations/unlike (where the analogies break down)

Protein synthesis is submicroscopic whereas building a house and baking bread are not

Blocks and ingredients for making bread can be cut up whereas AAs are always used in their entirely

A supervisor can make changes to a design or a chef can make changes to the cook instruction but in protein synthesis, no intention changes can be made

Proteins are naturally made by the body cells whereas bread baking/building a house is made/build by man.

Source: Reprinted with permission from Ref [42]. https://www.ijsciences.com/pub/article/1644

16.5 DISCUSSION

Working with NMs, the principle of caution should be applied. Research in 2D NMs augurs an electronics-world revolution. Keys for powerful, tiny-devices development are conceived in chemistry and physics laboratories throughout the world, end to which a molecular-interactions study is a tool. How does molecular science contribute to improving 2D-NMs properties, facing the development of future electronic devices?

274   Green Chemistry and Biodiversity: Principles, Techniques, and Correlations

Chalcogen NCs are analyte preconcentration, imaging, and detection tools from environmental to proteomic studies. People familiar with the matter and how to manipulate it created much civilization stuff.[33] They may not call themselves chemists but their experimentation shaped the modern world. For the sheer joy of knowing, alchemists, chemists, and molecular scientists sought to expand people's matter understanding from the atomic to the galactic.[34] Frontier science provides hope for problems, However, inequality persists. Science must contribute toward life being better than previous generation for everybody. To associate the word chemical with scientific laboratory is common, as something arising out of beakers and flasks at the hands of chemists wearing laboratory coats.[35] In 21st century, chemical stands juxtaposed to natural and, increasingly in public opinion, the two are considered mutually exclusive occurrences. While many manmade chemicals are toxic, the same occurs with plant products.[36] The artificial/natural dilemma comes from the times of alchemy: The cause is the lack of spreading.[37] The need for communicating science is a secondary effect of scientific creativity.

Gandía Balaguer tackled general bioethical aspects from historical notes to technique foundations, and concepts anthropological bioethics and law.[38] Subjects came unstrung [e.g., HGP (1990), cloning, stem cells, preimplantation/prenatal genetic diagnosis, gene therapy, human immunodeficiency virus/acquired immunodeficiency syndrome (HIV/AIDS), human-life end: Eldercare, death, suffering, euthanasia]. The ACS informed Academic Professional Guidelines. However, they do not explain what to do if the science policy comes into conflict with boss or employees. Medicine first challenge is to solve bioethical problems.[39] Bioethics subject is to look after life.[40] People never had as many possibilities of taking care of this as nowadays, on aspects related to factors: Hygiene, health, gynecology, pediatrics, geriatrics, transplants, palliatives, genetically modified, etc.[41] Notwithstanding, life was never as much threatened as these days: Wars, violence, unfair sharing goods, technology abuses, environmental destruction, etc. At the present time, people can look after life. However, they can destroy it with greater easiness.

Bioethics becomes ethics of responsible management of human life, in the framework of fast biomedical advances, and therapeutic limits or therapy/research frontiers must be rethought. However, laic ethics/religious perspectives meeting arises, where theological positions should abandon dogmatism, fundamentalism, and moral monopoly; in theological ethics,

Metaphors That Made History

one should make possible discrepancy without dissidence versus magisterial authorities, and tradition criticism with creative fidelity. Scientific questioning creates hypotheses, trusting discoveries. Fiduciary listening dares to wait that revelations become public. Both are capable of searching, questioning and changing one's mind faced with data, without allowing himself to be rubbed off by relativist-fear syndrome. Notwithstanding, ideologies, with science or faith mask, neither doubt nor investigate: They stick to deceptive certainties, longing for securities and creating pseudoscience/pseudoreligion, which are usually dogmatic-fanaticisms uniform.

16.6 FINAL REMARKS

From the present results and discussion, the following final remarks can be drawn.

(1) Chemistry is immersed in cultural values, which matter for the public acceptance of scientific and technological innovations. New trends show decaying differences: natural/social sciences, technoscience, physics/chemistry, economy/capitalism, religion/ philosophy, etc. Sign of neither prejudice nor enmity should exist in a literary work, nor researchers should be afraid of exceeding cultural limits in the search for truth or presenting objectively the viewpoint of the other.

(2) Science is seen as experimental, but another theoretical one exists and classification is part of a theory. Results bring chemical and physical properties into the flatland for their future use in applications. Chemistry is not a selfcontained discipline: It becomes nanochemistry while physics merges with materials science. The artificial/natural dilemma comes from the times of alchemy: The cause is lack of spreading. Need for communicating is a secondary effect of creativity. To create a critical conscience in the people who must speak of science is more important than to know science. The failure of risks-communication campaigns is journalism subject.

(3) People live in a fragile world and forget that economic growth is a means to a sustainable present of all humanity. Living on Earth, believing that no need for concepts of enough and equity exists, is risky. People should take the way of ethics based on resource

conservation, design and use of sustainable synthetic processes, waste reduction, human relationships, and global solidarity. Values should be the new world's roots.

(4) It remains to be solved the knowledge of the factors hindering the translation of science education research to teaching practice. As educators of the professionals of tomorrow, professors have the ethical duty of providing them an integral training, where teachers should include the need for protecting the environment. Professors invite educators to provide them solid knowledge, which allows them to take part in the struggle versus natural risks, and encourage predisposition to international collaboration and joint aid to the weakest/most exposed to catastrophes.

ACKNOWLEDGMENTS

The authors thank the support from Generalitat Valenciana (Project No. PROMETEO/2016/094) and Universidad Católica de Valencia *San Vicente Mártir* (Projects Nos. UCV.PRO.17-18.AIV.03 and 2019-217-001).

KEYWORDS

- **biomarker**
- **biosupervision**
- **culture**
- **deoxyribonucleic acid**
- **epidemiological study**
- **ethics**
- **medicine**

REFERENCES

1. Torrens, F.; Castellano, G. *Reflections on the Nature of the Periodic Table of the Elements: Implications in Chemical Education.* In *Synthetic Organic Chemistry*; Seijas, J. A., Vázquez Tato, M. P., Lin, S. K., Eds.; MDPI: Basel, Switzerland, 2015; Vol. 18; pp 1–15.

2. Torrens, F.; Castellano, G. *Nanoscience: From a Two-Dimensional to a Three-Dimensional Periodic Table of the Elements.* In *Methodologies and Applications for Analytical and Physical Chemistry*; Haghi, A. K., Thomas, S., Palit, S., Main, P., Eds.; Apple Academic–CRC: Waretown, NJ, 2018; pp 3–26.

3. Torrens, F.; Castellano, G. *Periodic Table.* In: *New Frontiers in Nanochemistry: Concepts, Theories, and Trends*; Putz, M. V., Ed.; Apple Academic–CRC: Waretown, NJ, in press.

4. Torrens, F.; Castellano, G. *Ideas in the History of Nano/Miniaturization and (Quantum) Simulators: Feynman, Education and Research Reorientation in Translational Science.* In *Synthetic Organic Chemistry*; Seijas, J. A., Vázquez Tato, M. P., Lin, S. K., Eds.; MDPI: Basel, Switzerland, 2015; Vol. 19; pp 1–16.

5. Torrens, F.; Castellano, G. *Reflections on the Cultural History of Nanominiaturization and Quantum Simulators (Computers).* In *Sensors and Molecular Recognition*; Laguarda Miró, N., Masot Peris, R., Brun Sánchez, E., Eds.; Universidad Politécnica de Valencia: València, Spain, 2015; Vol. 9; pp 1–7.

6. Torrens, F.; Castellano, G. *Nanominiaturization and Quantum Computing.* In *Sensors and Molecular Recognition*; Costero Nieto, A. M., Parra Álvarez, M., Gaviña Costero, P., Gil Grau, S., Eds.; Universitat de València: València, Spain, 2016; Vol. 10 pp 31-1–31-5.

7. Torrens, F.; Castellano, G. *Nanominiaturization, Classical/Quantum Computers/Simulators, Superconductivity, and Universe.* In *Methodologies and Applications for Analytical and Physical Chemistry*; Haghi, A. K., Thomas, S., Palit, S., Main, P., Eds.; Apple Academic–CRC: Waretown, NJ, 2018; pp 27–44.

8. Torrens, F.; Castellano, G. *Superconductors, Superconductivity, BCS Theory and Entangled Photons for Quantum Computing.* In *Physical Chemistry for Engineering and Applied Sciences: Theoretical and Methodological Implication*; Haghi, A. K., Aguilar, C. N., Thomas, S., Praveen, K. M., Eds.; Apple Academic–CRC: Waretown, NJ, 2018; pp 379–387.

9. Torrens, F.; Castellano, G. *EPR Paradox, Quantum Decoherence, Qubits, Goals, and Opportunities in Quantum Simulation.* In *Theoretical Models and Experimental Approaches in Physical Chemistry: Research Methodology and Practical Methods*; Haghi, A. K., Ed.; Apple Academic–CRC: Waretown, NJ, 2018; Vol. 5, pp 317–334.

10. Torrens, F.; Castellano, G. *Nanomaterials, Molecular Ion Magnets, Ultrastrong and Spin–Orbit Couplings in Quantum Materials.* In: *Physical Chemistry for Chemists and Chemical Engineers: Multidisciplinary Research Perspectives*; Vakhrushev, A. V., Haghi, R., de Julián-Ortiz, J. V., Allahyari, E., Eds.; Apple Academic–CRC: Waretown, NJ, in press.

11. Torrens, F.; Castellano, G. *Nanodevices and Organization of Single Ion Magnets and Spin Qubits.* In *Chemical Science and Engineering Technology: Perspectives on Interdisciplinary Research*; Balköse, D., Ribeiro, A. C. F., Haghi, A. K., Ameta, S. C., Chakraborty, T., Eds.; Apple Academic–CRC: Waretown, NJ, in press.

12. Torrens, F.; Castellano, G. *Superconductivity and Quantum Computing* via *Magnetic Molecules.* In *New Insights in Chemical Engineering and Computational Chemistry*; Haghi, A. K., Ed.; Apple Academic–CRC: Waretown, NJ, in press.

13. Torrens, F.; Castellano, G. *Developing Sustainability* via *Nanosystems and Devices: Science–Ethics.* In *Chemical Science and Engineering Technology: Perspectives on*

278 Green Chemistry and Biodiversity: Principles, Techniques, and Correlations

Interdisciplinary Research; Balköse, D., Ribeiro, A. C. F., Haghi, A. K., Ameta, S. C., Chakraborty, T., Eds.; Apple Academic–CRC: Waretown, NJ, in press.

14. Torrens, F.; Castellano, G. Green Nanotechnology: *An Approach towards Environment Safety*. In *Advances in Nanotechnology and the Environmental Sciences: Applications, Innovations, and Visions for the Future*; Vakhrushev, A. V.; Ameta, S. C.; Susanto, H., Haghi, A. K., Eds.; Apple Academic–CRC: Waretown, NJ, in press.

15. Torrens, F.; Castellano, G. *Molecular Devices/Machines: Hybrid Organic–Inorganic Structures*. In *Research Methods and Applications in Chemical and Biological Engineering*; Pourhashemi, A., Deka, S. C., Haghi, A. K., Eds.; Apple Academic–CRC: Waretown, NJ, in press.

16. Torrens, F.; Castellano, G. *The Periodic Table, Quantum Biting its Tail, and Sustainable Chemistry*. In *Chemical Nanoscience and Nanotechnology: New Materials and Modern Techniques*; Torrens, F., Haghi, A. K., Chakraborty, T., Eds.; Apple Academic–CRC: Waretown, NJ, in press.

17. Torrens, F.; Castellano, G. *Quantum Molecular Spintronics, Nanoscience, and Graphenes*. In *Molecular Physical Chemistry*; Haghi, A. K., Ed.; Apple Academic–CRC: Waretown, NJ, in press.

18. Torrens, F.; Castellano, G. *Cancer and Hypotheses on Cancer*. In *Molecular Chemistry and Biomolecular Engineering: Integrating Theory and Research with Practice*; Pogliani, L., Torrens, F., Haghi, A. K., Eds.; Apple Academic–CRC: Waretown, NJ, in press.

19. Torrens, F.; Castellano, G. *Precision Personalized Medicine from Theory to Practice: Cancer*. In *Molecular Physical Chemistry*; Haghi, A. K., Ed.; Apple Academic–CRC: Waretown, NJ, in press.

20. Torrens, F.; Castellano, G. AIDS destroys immune defences: Hypothesis. *New Front. Chem.* 2014, 23, 11–20.

21. Torrens-Zaragozá, F.; Castellano-Estornell, G. Emergence, spread and uncontrolled Ebola outbreak. *Basic Clin. Pharmacol. Toxicol.* **2015,** *117* (Suppl. 2), 38–38.

22. Torrens, F.; Castellano, G. 2014 spread/uncontrolled Ebola outbreak. *New Front. Chem.* 2015, *24*, 81–91.

23. Torrens, F.; Castellano, G. Ebola virus disease: Questions, ideas, hypotheses, and models. *Pharmaceuticals* **2016,** *9*, 14–6-6.

24. Strauss, L. *El Gusto de Jenofonte: Una Introducción a la Filosofía*; Biblioteca Nueva: Madrid, Spain, 2018.

25. Pascual, S., Ed. *Per què Filosofia?* Neopàtria: Alzira, València, Spain, 2017.

26. Kukso, F. La nueva revolución genómica de CRISPR hace florecer las metáforas. *Anuario SINC* **2016,** *2016*, 120–123.

27. Keller, E. F. *Refiguring Life: Metaphors of Twentieth-Century Biology*; Columbia University: New York, NY, 1996.

28. Lakoff, G.; Johnson, M. *Metaphors We Live by*; University of Chicago; London, UK, 2003.

29. Sontag, S. *Illness as Metaphor*; Farrar, Straus and Giroux: New York, NY, 1978.

30. Lizcano, E. *Metáforas Que nos Piensan: Sobre Ciencia, Democracia y Otras Poderosas Ficciones*; Bajo Cero–Traficantes de Sueños: Madrid, Spain, 2006.

31. Baldridge, K.; Abendroth, T. *Mixing with MerryGold: A Faity Tale*; Wiley–VCH: Weinheim, Germany, 2008.

32. Sánchez Ruiz, J. ¡SuperJ al recate! La investigación en clave de cómic: Por qué no hay aún una *cura* contra el *cáncer*? *Cris T Cuenta (Madrid)* **2017,** *2017* (8), 8–8.

33. Bunge, M. *La Investigación Científica: Su Estrategia y su Filosofía*; Ariel: Barcelona, Spain, 1985.

34. Simonyi, K. *A Cultural History of Physics*; CRC: Boca Raton, FL, 2012.

35. Boudry, M., Pigliucci, M., Eds. *Science Unlimited? The Challenges of Scientism*; University of Chicago: Chicago, IL, 2017.

36. Johns, T. *With Bitter Herbs They Shall Eat: Chemical Ecology and the Origins of Human Diet and Medicine*; University of Arizona: Tucson, AZ, 1990.

37. Bensuade-Vincent, B.; Simon, J. *Chemistry: The Impure Science*. Imperial College: London, UK, 2012.

38. Gandía Balaguer, A. *Bioética y Ciencia: Monográfico de Bioética*; Universidad Católica de Valencia *San Vicente Mártir*: València, Spain, 2006.

39. Salvi, M. *Bioética en Europa: Diferencias, Coincidencias e Iniciativas Políticas*; Humanitas, Humanidades Médicas No. 44. Fundación Medicina y Humanidades Médicas: Barcelona, Spain, 2009.

40. Masià, J. *Cuidar la Vida: Debates Bioéticos*; Herder: Barcelona, Spain, 2012.

41. Trullenque Peris, R. *L'Estat Actual de la Medicina: Perspectives de Futur*; Arxius i Documents No. 66. Institució Alfons el Magnànim–CVEI–Diputació de València: València, Spain, 2017.

42. Ameyaw, Y.; Kyere, I. Analogy-based Instructional Approach a Panacea to Students' Performance in Studying Deoxyribonucleic Acid (DNA) Concepts. *Int. J. Sci.* **2018,** *7* (5), 7–13.

43. Gick, M. L.; Holyoak, K. J. Analogical Problem Solving. *Cog. Psychol.* **1980,** *12,* 306–355.

Index

A

Aegle marmelos (Bael), 35–36
  beneficial parts, 36
  biochemical and biological aspects, 38–39
    abundant in phytochemicals, 39
    nutrients, 38
    rich in antioxidants, 38–39
  classification, 37
  different names, 37
  traditional knowledge, 40
  uses, 40–44
    diarrhea, 42
    diuretic activity, 43
    gastric ulcers, 41
    high cholesterol and triglyceride levels, 41
    hypertension, 43
    immunomodulatory potential, 44
    inflammatory bowel disease, 42–43
    liver injury, 42
    type 2 diabetes mellitus, 40–41
Aloe vera
  phytopharmacotherapy, 14
American Chemical Society (ACS), 269
Anthropogenic contaminants, 2

B

Bilva, 35–36
  beneficial parts, 36
  biochemical and biological aspects, 38–39
    abundant in phytochemicals, 39
    nutrients, 38
    rich in antioxidants, 38–39
  classification, 37
  different names, 37
  uses, 40–44
    diarrhea, 42
    diuretic activity, 43
    gastric ulcers, 41
    high cholesterol and triglyceride levels, 41
    hypertension, 43
    immunomodulatory potential, 44
    inflammatory bowel disease, 42–43
    liver injury, 42
    type 2 diabetes mellitus, 40–41
Biodiversity, 23–24
  conservation strategies
    control emerging threats, 29–31
    maintain intact (viable) landscapes, 29
    recover threatened species, 29
    reverse declines, 29
  elements, 25–27
  loss, causes, 27–29
  measurement, 24
Biosphere reserves, 31
Black tea leaves, 163
Bone disorders, 17–18
Bone health
  natural products, 14–15
  discussion, 15–17
Bottom up approach, 47
  AuNPs, 47–48
Brazilian biodiversity, 9, 11–12, 17–18
  source of new drugs, 11–12
Breast cancer (BC), 209

C

Carbolines, 137
Carnation *(Dianthus caryophyllus)*, 137–138
  experimental
    determination of tannins content, 140–141
    DPPH assay, 142

preparation of stock and working solutions, 140
rotational viscometry, 141
flower petals, 138
results and discussion, 143–148
IC_{50} values, 146
percentage of reduction of ABTS, 146–148
time dependent changes in dynamic viscosity, 143–145
therapeutic benefits, 138
therapeutic benefits of oil, 138
Central nervous system (CNS), 244
Chalcogen nanocomposites (NCs), 268
Ciprofloxacin dimers, 244
Citrus peel extracts, 158
Colloidal stability, 53
Community or ecosystem diversity, 25–27
Compositional diversity, 25
Conservation strategies of biodiversity
control emerging threats, 29–31
biosphere reserves, 31
national parks, 30
protected areas, 30
sanctuaries, 30
maintain intact (viable) landscapes, 29
recover threatened species, 29
reverse declines, 29
Conventional cuboctahedral structure, 57
Cordia verbenacea, 9–10
Curcumin, 169–170
bioapplicability
antibacterial activity with curcumin-loaded HSNC, 180
curcumin loading and release studies, 179
in vitro curcumin release and loading, 179–181
discussion, 171–178
Au nanorod formation, 175
chemical structure, 173
deprotonation, 174
effect of pH, 172–175
effect of temperature, 177–178
pH effect on drug loading, 175–176
release of curcumin drug by INVITE micelles, 177

and releasing with PDEA block, 175–176
reprinted with permission, 172
solubility of curcumin, 172
TEM images, 177
in vitro curcumin release study, 176–177
medicinal properties, 170
oral administration to Tg2576 AD model, 171
PDEA blocks, 171
structure, 170
Cur/MPEGPCL28 micelles, 181
Cyanuric chloride, 244
Cycas leaves, 163–164

D

Dianthus caryophyllus, 135–136
2, 4-dimethoxy-6-(piperazin-1-yl)-1,3,5-triazine aryl substituted ureido/ aryl amido derivatives, 245, 247
Diosmin
designing nanometals for, 155
transmission electron microscopy, 155–156

E

Echinacea purpurea, 9–10
triple way of action, 13
Echinaforce®, 17
Ellagic acid (EA), 111, 113, 115–116
bioavailability, 123–127
effectiveness of nutraceuticals, 124
poor absorption, 124–125
urolithins production, 126
bioprocess to produce, 117–119
production by SSC, 119
SSC, 117–118
tannase, 119
biotransformation process of ETs, 113
importance, 113
industrial uses, 123
molecule, 115
natural compounds, 116–117
evidence of presence, 116
sources, 116–117

Index

283

properties and application, 119–123
  antimicrobial activity, 121–122
  antimutagenic and anticarcinogenic activity, 120
  antioxidant activity, 120–121
  antiviral activity, 122–123
Ellagitannins (ETs), 114–115
Endocrine-disrupting chemicals (EDCs), 2
  exposure, 3
Environmental contamination, 2
Environmental engineering science, 188
  nanotechnology applications, 189
Environmental sustainability, 190–191
Essential oil (EO), 10

F

Flavonoids, 153–154
  biosynthesis, 165
  designing nanometals for, 155–165
    $AgNP_s$ and $AuNP_s$, 160–161
    antimicrobial activity of synthesized NPs, 157
    bioinspired synthesis $AuNP_s$, 159–160
    black tea leaves, 163
    crystalline nature of $HtAuNP_s$, 159
    cycas leaves, 163–164
    nanoformulated insecticides and antimicrobials, 164–165
    plant leaf extracts, 162
    TEM images, 161–162
    transmission electron microscopy, 155–156
  quercetin, 160
  structure, 155
Functional diversity, 25

G

Gallotannins and ellagitannins, 82
Genetic diversity, 25–26
Global water crisis, 188
Gold nanoparticles (AuNPs), 47–48
Grape (*Vitis vinifera* L.), 138–139
  commercial preparations, 139
  experimental
    ABTS assay, 142

    determination of tannins content, 140–141
    DPPH assay, 142
    preparation of stock and working solutions, 140
    rotational viscometry, 141
  material and methods
    chemicals, 139–140
  results and discussion, 143–148
    IC_{50} values, 146
    percentage of reduction of ABTS, 146–148
    time dependent changes in dynamic viscosity, 143–145
  seeds, 139

H

Herbaceous ecosystems (biogeocenoses), 68
Herd animals, 69
Hesperidin
  designing nanometals for, 155
  transmission electron microscopy, 155–156
Heterocyclic compounds, 243
  piperazine, 244
Human factor engineering, 193

I

Integrated water resource management, 189–190, 193
Isobutylamides, 16

L

Larrea, chaparral, 98
Larrea tridentata, 97
  experimental of polyphenolic compounds
    antioxidant activity assay, 101–102
    condensed tannins assay, 101
    HPLC/MS assay, 102
    raw material, 99–100
    reactive and standards, 100
    reflux extraction, 100
    total phenolic compounds assay, 101

ultrasound extraction, 100–101
results and discussions, 102–107
  characterized compounds in HPLC/
    MS, 106–107
  condensed tannins graph, 104
  % DPPH radical of reflux, 102–103
  HPLC/MS analysis, 105
  total polyphenol hydrolyzable graph,
    103
  ultrasound-assisted, 102–103
ultrasonic extraction, 97–98
Loss of biodiversity
  causes, 27–29
  climate change, 27–28
  habitat change, 27
  invasive alien species, 28
  overexploitation, 28
  poaching, 28–29
  pollution and global warming, 28

M

Meadow and field weeds, 65–66
  materials and methodology, 67
  results and discussion, 67–76
Medicinal plants, 18, 65–66
  biologically active substances, 65
  weeds, 66
Membrane science and novel separation
  processes
  vision of, 190
Metal nanoparticles, 47
Metaphors
  bioethical aspects, 269
  cancer description, 271
  clustered, 270
  DNA, 270
  history in science, 271
  house/bread-baking
    analogies, 272–273
  Human Genome Project (HGP, 1990),
    270
  philosophy, 270
  regularly interspaced, 270
  short palindromic repeat (CRISPR), 270
Microfiltration, 192–193

N

Nanocurcumin (NC), 181
Nanofiltration, 192–193
  significant scientific research pursuit,
    198–200
Nanoformulated insecticides and
  antimicrobials, 164–165
Nanotechnology, 193
  significant scientific research pursuit,
    198–200
Naringin
  designing nanometals for, 155
  transmission electron microscopy,
    155–156
Natural antioxidants, 11
Natural compounds, 116–117
  EA sources, 116–117
  evidence of presence of EA, 116
Novel piperidine
  anti-inflammatory assay
    antibacterial assay, 258
    antifungal assay, 258–259
    lip polysaccharide (LPS), 257
    structure activity relationships (SAR),
      259–261
  biological evaluation
    cytokines, 256
    interleukin-6 (IL-6), 256
    triazine, 257
    tumor necrosis factor alpha (TNF-α),
      256–257
  chemicals and solvents, 247
  Ciprofloxacin dimers, 244
  cyanuric chloride, 244
  design scaffold, 246
  2, 4-dimethoxy-6-(piperazin-1-yl)-1,3,5-
    triazine aryl substituted ureido/ aryl
    amido derivatives, 245, 247
  4-(4,6-dimethoxy-1,3,5-triazin-2-YL)
    piperazin-1-YL) (substituted phenyl)
    methanone amide derivatives (3I-P)
    antibacterial activity, 263
    antifungal activity, 264
    anti-inflammatory activity data, 263

Index 285

4-(4,6-dimethoxy-1,3,5-triazin-2-YL)
  piperazin-1-YL) (substituted phenyl)
  methanone amide derivatives (3I–P)
  analytical and mass spectral data,
    253–254
  ^{1}H NMR data, 255
  IR spectral data, 254
  procedure, 253
4-(4,6-dimethoxy-1,3,5-triazin-2-YL)-
  N-substituted phenylpiperazine-1-
  carboxamide urea derivatives (3A–H)
  analytical and mass spectral data,
    250–251
  antifungal activity, 262
  anti-inflammatory activity data, 261
  IR spectral data, 251
  procedure, 249–250
heterocyclic compounds, 243
  piperazine, 244
IR spectra, 247
N-substituted piperazine annulated
  s-triazine derivatives, 245
physicochemical data, 248
  procedure, 249
piperazine, 244
tert-butyl 4-(4,6-dimethoxy-1,3,5-
  triazin-2-YL) piperazine-1-
  carboxylate (3C), synthesis
  procedure, 248
trisubstituted s-triazines, 245
N-substituted piperazine annulated
  s-triazine derivatives, 245

P

Pasture erosion, 70
  depression of productivity, 70–71
Peganum harmala, 135–136
Perennial grasses
  role in protection of soil, 68–69
Periodic table of elements (PTE), 269
Phenolic compounds, 82
Photocatalysis, 2
Phytochemicals, 136
Piperazine, 244
Plant species from tropical forest, 10
Polyphenolic compounds

extraction, 99
Polyphenolic phytochemicals, 11
Polyphenols, 98–99, 112–113
  importance, 114
Polyphenols-rich plant extracts
  discussion, 92
  materials and methods
    acetylation, 84–85
    crude extract fractionation, 84
    extract preparation, 83–84
    gas cromatography–mass
      spectrometry analysis, 85
    hydrolysis, 84–85
    methylation, 84–85
    physicochemical characterization, 83
    plant material, 83
    sugar standard preparation, 84
    and trisilation, 84–85
  potency, 81
  results
    proximal chemical composition,
      85–86
    sugar analysis from *E. Antisyphilitica,*
      86–88
    sugar analysis from *J. dioica,* 89–91
    sugar analysis from *L. Tridentata,*
      88–89
Pomegranate (*Punica granatum* L.), 136
  beneficial effect, 136
  experimental
    determination of tannins content,
      140–141
    DPPH assay, 142
    preparation of stock and working
      solutions, 140
    rotational viscometry, 141
  results and discussion, 143–148
    IC_{50} values, 146
    percentage of reduction of ABTS,
      146–148
    time dependent changes in dynamic
      viscosity, 143–145
PPM oncology (PO), 209
Precision personalized medicine (PPM), 209
  5-amino-2-aroylquinolines, 210
  4-aroyl-6,7,8-trimethoxyquinolines, 210
  clinical research in, 225–230

286   Index

molecular biology to medicine, 211–220
present and future, 220–225
superJ rescue! comic-style research,
230–231
Punica granatum, 135–136

Q

Quercetin, 160

R

Reflections, 268

S

Seed growth method, 47–49
alkylthiols, 48
AuNPs role, 48
Brust–Schiffrin method modification, 49
effect of
ascorbic acid, concentration, 55–56
mode of addition of reducing agent,
49–52
surfactant, concentration, 56–57
electron micrographs of Au particles, 50
generalized two-step mechanism, 51
role of AG$^+$ ION, 57–60
Turkevich synthesis, 52–55
SnO$_2$ quantum dots (QDs), 1
experimental
comparative typical run of TiO$_2$ and
TiO2/SnO$_2$, 6
degradation profile of 2-nitrophenol,
4–7
preparation, 4
typical run, 5–6
Species diversity, 25–26
Structural diversity, 25
Surfactants, 56
concentration, 56–57
Syrian rue *(Peganum harmala L.),* 137
experimental
ABTS assay, 142
determination of tannins content,
140–141
DPPH assay, 142
preparation of stock and working
solutions, 140

rotational viscometry, 141
results and discussion, 143–148
IC$_{50}$ values, 146
percentage of reduction of ABTS,
146–148
time dependent changes in dynamic
viscosity, 143–145

T

Tabernaemontana elegans leaves
bioactivities of indole alkaloids, 13–14
monoterpene and β-carboline indole, 14
Tanin-acyl hydrolase (tannase), 119
Tannins, 82, 112–113
importance, 114
medicinal properties, 82
structural characteristics, 82
Top up approach, 47
Transmission electron microscopy,
155–156
Trisubstituted s-triazines, 245
Turkevich method/Turkevich synthesis,
52–55
steps in process, 52–53

U

Ultrafiltration, 192–193
Ultrasound-assisted extraction, 97–98
Urolithins, 113–114, 125–126
production from EA, 126

V

Variability, 23–24
Vegetal-drugs preparations, 17–18
Vitis vinifera, 135–136

W

Water purification, 187
arsenic, 191–192
heavy metal groundwater remediation,
191–192
nanotechnology, 192
scientific advances in, 193–198
Water resource management, 188